Emerging Technologies in Clean Energy for the Twenty-First Century 2015

Topical Conference at the 2015 AIChE Spring Meeting and 11th Global Congress on Process Safety

Austin, Texas, USA
26-30 April 2015

ISBN: 978-1-5108-0693-1

Printed from e-media with permission by:

Curran Associates, Inc.
57 Morehouse Lane
Red Hook, NY 12571

CURRAN ASSOCIATES INC.
proceedings
.com

Some format issues inherent in the e-media version may also appear in this print version.

Printed by Curran Associates, Inc. (2015)

For permission requests, please contact AIChE
at the address below.

AIChE
3 Park Avenue
New York, NY 10016-5991

Phone: (203) 702-7660
Fax: (203) 775-5177

www.aiche.org

Additional copies of this publication are available from:

Curran Associates, Inc.
57 Morehouse Lane
Red Hook, NY 12571 USA
Phone: 845-758-0400
Fax: 845-758-2634
Email: curran@proceedings.com
Web: www.proceedings.com

TABLE OF CONTENTS

409505 Chemical Engineering: Is This the Golden Age?

Abstract:

Schlanger will offer his insights on the current opportunities in the chemical process industries in his talk entitled "Chemical Engineering: Is This the Golden Age?" on Monday morning, April 27. At the conclusion of his lecture, Schlanger will receive the AGILE Award in recognition of his leadership at CEVA Group, a $9 billion logistics supplier to industry worldwide; LyondellBasell, a major player in the plastics, chemicals, and refining industries; Arco Chemical; and at the other companies where he has held leadership positions.

400449 Effect of Ozone Concentrations from Local Emission Sources

Abstract:

Ground level ozone is one of six criteria air pollutants designated by US EPA. It is not emitted directly into the air, but is created by chemical reactions between oxides of nitrogen (NOx) and volatile organic compounds (VOC) in the presence of sunlight. Emissions from industrial facilities and electric utilities, motor vehicle exhaust, gasoline vapors, and chemical solvents are some of the major sources of NOx and VOC.

In this report, rigorous air quality simulation was performed to simulate the ozone formation of the Beaumont & Port Arthur, Texas (BPA) area. Simulation results match observed data from monitoring stations. Ozone source apportionment showed local emission sources contributed 20% of total ozone. For the local contributions, point souces such as industrial facilities and electric utilities contributed 10% of total ozone, and on-road vehicles contributed 5%. Matrix simulations and EKMA diagrams were also performed to show the effects of emission reductions of anthropogenic NOx and VOC sources.

Emissions Testing of Sonic Velocity Flares Validates High Destruction and Removal Efficiency

Scot Smith
Director of Flare Products
Zeeco, Inc.
22151 East 91st Street
Broken Arrow, Okla. 74014
Scot_smith@zeeco.com

ID #400369

Prepared for Presentation at
American Institute of Chemical Engineers
2015 Spring Meeting
11th Global Congress on Process Safety
Austin, Texas
April 27-30, 2015

UNPUBLISHED

Abstract

In a 2012 report by the EPA's Flare Review Panel, regulations for flares that tended to operate outside their stable flame envelope were expanded to include limits on velocity as a function of net heating value.

EPA regulations do not currently allow sonic velocity flares to be permitted and operated without first performing an alternate means of emissions limitation test to validate destruction efficiencies. Existing regulations affect the application of multi-point staged flares since these flares operate at sonic exit velocities that exceed the maximum exit velocity requirements cited in 40 CFR 60.18 and 40 CFR 63.11(b) and applicable state regulations.

Over the years, Zeeco has conducted numerous destruction efficiency tests on its multi-point sonic flares. All of these tests demonstrated high destruction efficiency at velocities exceeding the EPA's limits. The purpose of these tests, performed at Zeeco's industrial-scale combustion test facility near Tulsa, Oklahoma, was to validate sonic tips for more applications using an expanded range of gases and conditions. Extractive sampling was used to measure the emissions from the exhaust plume at staging and de-staging pressures. The extracted gas concentrations were used to calculate the destruction efficiency of the multi-point sonic flare tips at each test point. Even with sonic gas exit velocities, the Zeeco-designed sonic flare tips maintained high destruction efficiency.

Despite pressure from both end users and regulators, Zeeco's flare burner tests and results have been maintained internally. This paper will present those results to the industry for the first time.

Benefits and Characteristics of Multi-Point Ground Flares

Multi-Point Ground Flares (MPGF) derive their name from their physical layout. Typically, they are a field of multiple pressure-assisted flare tips that are vertically mounted at grade and arranged in stages that open as the upstream pressure (gas flow) increases, and close as pressure decreases. This design concept is the same today as it was when MPGF were invented over 40 years ago. MPGF are often selected to attain smokeless combustion for heavy hydrocarbon service with high available pressure, and have proven to provide stable operation at sonic velocities for a wide range of gas compositions. MPGF are also used in situations where the designer wants to reduce or eliminate radiation or visible flame. The high pressure is used to assist the gas to obtain smokeless operation for the full range of flaring capacity, which can be difficult to do with other assist mediums such as air, gas, or steam due to high smokeless rates. Each burner has unobstructed air access allowing the momentum from the high exit velocity of the flare gas to entrain the necessary air for combustion. MPGF are designed to provide maximum smokeless performance, while minimizing radiation impacts and plot space. Installing a fence around the field can block the visibility of the flame; this serves a dual purpose by reducing radiation outside the fenced area and reducing the likelihood that flaring operations will be a nuisance to the public. The figure below is an example of a typical MPGF installation.

Figure 1 – Typical Installation of a MPGF

Another benefit of MPGF is easy access for maintenance, because all of the staging equipment is located at grade and outside of the fence. As a result, personnel can access the staging equipment safely without being affected by a flaring event. Figure 2 illustrates the staging equipment and the exterior of a MPGF fence.

Figure 2 – Typical Installation of MPGF Fence

Over the past 40-plus years, there have been many improvements in burner technology. Zeeco's modern pressure-assisted tips are custom designed to optimize the mixing of flare gas and air thus improving tip performance. Zeeco's pressure-assisted flare tips, shown in Figure 3, are an investment casting typically made out of 310 SS casting material.

Figure 3 – Zeeco's Pressure-Assisted Flare Tip Casting

Multiple tips are arranged in stages to allow control over the number of tips in service depending upon the pressure and flow rate of the flare gas. This enables the tips to operate in the optimal pressure range for maximum smokeless capacity. Each stage will open when the pressure in the MPGF header has increased to the designed staging pressure and subsequently close when the de-stage pressure is met. Figure 4 is an example of a typical staging curve used to control MPGF.

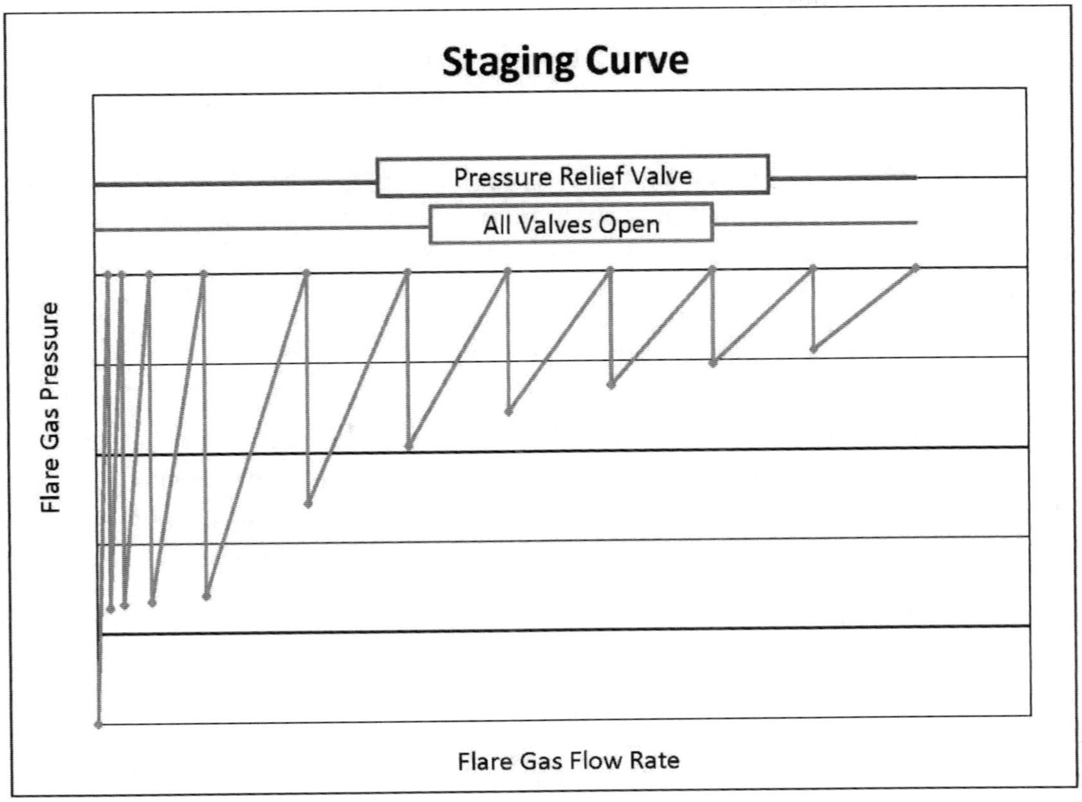

Figure 4 – Typical Staging Curve for MPGF

The staggered blue line in Figure 4 represents how the pressure in the system will change as each stage opens and closes. The high and low points of the blue line represents the staging and de-staging pressures for each stage, respectively. Typically, the staging pressure is equal to the available pressure of the system; however, it could be lower based on maintaining stable performance. When the pressure in the MPGF header reaches the designed staging pressure, the next stage of tips opens. The system is designed so that the pressure in the open stages never decreases below a desired pressure. This is referred to as the de-staging pressure and generally corresponds to the minimum pressure required to ensure stable performance and smokeless combustion. A typical de-staging curve follows the same pattern as the staging curve but closes the valves as the pressure in the MPGF header decreases.

Regulations and Industry Testing

Currently, federal and state regulations on flares limit the exit velocity of the flare gas based on the composition and LHV of the gas as well as the assist medium of the flare. There are three assist types considered by the regulations: non-assisted, steam-assisted, and air-assisted. According to the federal and state regulations, if a flare is operated within these velocity and LHV restrictions, a destruction and removal efficiency (DRE) of 98% or greater is guaranteed. This suggests that if a flare operates outside of these limits, it will result in lower DRE.

In 1983, flare efficiency testing was performed by the Chemical Manufacturers Association (CMA). Three types of flares were tested: air-assisted, steam-assisted, and non-assisted. It is important to note that pressure-assisted flares were not included in this testing. In the testing performed, extractive sampling was used to measure the concentrations of emissions from each flare to determine the combustion efficiency (CE). The results from this testing became the basis of the current regulations: 40 CFR 60.18 & 40 CFR 63.11. The CMA concluded if there was a stable flame, the flare had high CE.

From 1984-86, Energy and Environmental Research (EER) Corporation performed testing for the EPA to evaluate the combustion efficiency CE for a variety of gas compositions on different commercially available flare tips, including two pressure-assisted tips that were obtained from flare vendors. The pressure-assisted flare tips were referred to in the test report as Commercial Tips "E" and "F." Commercial Tip E used a horizontal bar geometry while Commercial Tip F used an open geometry. The results showed that stable burning pressure-assisted tips operated at CE values of greater than 98%, even at high exit velocities.

In November 2013, Dow Chemical Company performed an Alternate Means of Emissions Limitation (AMEL) test for two of its MPGF installations. Nominal four-inch sonic burners were tested by extractive sampling to investigate the propylene DRE of two different flare tips, one of which was pressure-assisted. The results from this testing showed that the pressure-assisted tips were capable of providing greater than 98% DRE and CE over the range of gases that were tested. Furthermore, the testing confirmed that the presence of a stable flame ensured high DRE and CE. The conclusions and results were published and presented at the American Flame Research Committee (AFRC) Conference in 2014.

The current EPA regulations are based on the CMA testing in 1983; however, various other tests have been performed to further investigate pressure-assisted flare efficiency. These tests prove sonic flares can operate outside of the limits set by the EPA and still achieve high DRE.

Execution of Extractive Sampling

In addition to the testing previously described, Zeeco has conducted numerous destruction and removal efficiency tests on its MPGF pressure-assisted tips. All MPGF testing that is further detailed in this paper was conducted at the Zeeco Inc. Combustion and Research Test Facility in Broken Arrow, Oklahoma. The purpose of Zeeco's tests on MPGF was to investigate the destruction and removal efficiency and stability over a wide range of net heating values (NHV) and gas compositions. NHVs from 440 BTU/SCF to 2316 BTU/SCF were tested with gas mixtures including the following: propane, propylene, natural gas, carbon dioxide, nitrogen, and hydrogen. Destruction and removal efficiency was determined by measuring the concentration of carbon dioxide (CO_2), carbon monoxide (CO), and total hydrocarbons (THC) from extractive sampling. An example of extractive sampling being performed on a single MPGF pressure- assisted tip is shown in Figure 5.

Figure 5 – Extractive Sampling Method Used for Testing

Zeeco's custom sample hood, shown in Figure 5, was used to extract combustion products from the plume. A Venturi nozzle located at the end of the apparatus created a vacuum for sample extraction. A crane was used to position the inlet of the sample hood in the plume of combustion products. The crane was directed by an operator who monitored the hood's position with respect to the plume using a forward looking infrared (FLIR) camera to ensure a proper sample was obtained. A continuous sample was pulled from the sample hood through a heated sample line to gas analyzers where the concentration of the combustion products was measured and recorded.

The concentrations of CO_2, CO, and THC were used to calculate destruction and removal and combustion efficiency. DRE and CE are two ways of quantifying the degree of completion of the combustion reaction based on measured emissions. DRE is how well a component of interest is destroyed or broken down according to the amount of unburned hydrocarbons after the combustion process is completed. Alternatively, CE is how well a component of interest is converted into CO_2 and H_2O after the combustion process. The hydrocarbon DRE and CE equations are shown below.

$$DRE_{THC}(\%) = \frac{CO_2 + CO}{CO_2 + CO + THC} \cdot 100\%$$

$$CE(\%) = \frac{CO_2}{CO_2 + CO + THC} \cdot 100\%$$

It is important to understand the difference between both values to properly interpret the results. The CE is equal to or less than the DRE since the component of interest may be destroyed, but not completely combusted. The component of interest can reduce to an intermediate combustion product instead of completely combusting and forming only CO_2 and H_2O. To summarize, generating proportionally high CO_2 will result in high CE, while generating proportionally low hydrocarbons will result in high DRE. For conservative results, the DRE and

CE of total hydrocarbons were observed to verify that all of the hydrocarbons were being combusted.

Results

The following results are from numerous tests performed at Zeeco from 2013 to 2015. Gas mixtures with NHVs ranging from 440 to 2316 BTU/SCF were tested on a single piloted sonic flare tip to investigate the effect of NHV on flame stability and efficiency. The maximum allowable exit velocity per 40 CFR 60.18 is based on the NHV of the flare gas. For NHVs ranging from 200 BTU/SCF to 1000 BTU/SCF, an exponential function limits the exit velocity from 45.7 ft/s to 400 ft/s. For NHVs greater than 1000 BTU/SCF, the maximum allowable exit velocity remains 400 ft/s. Every gas was tested at exit velocities that exceeded the limits defined by 40 CFR 60.18. As long as the flame was stable, DRE and CE values greater than 99% were consistently observed. Figure 6 illustrates the high destruction and removal efficiencies obtained outside of the 40 CFR 60.18 operating limitations.

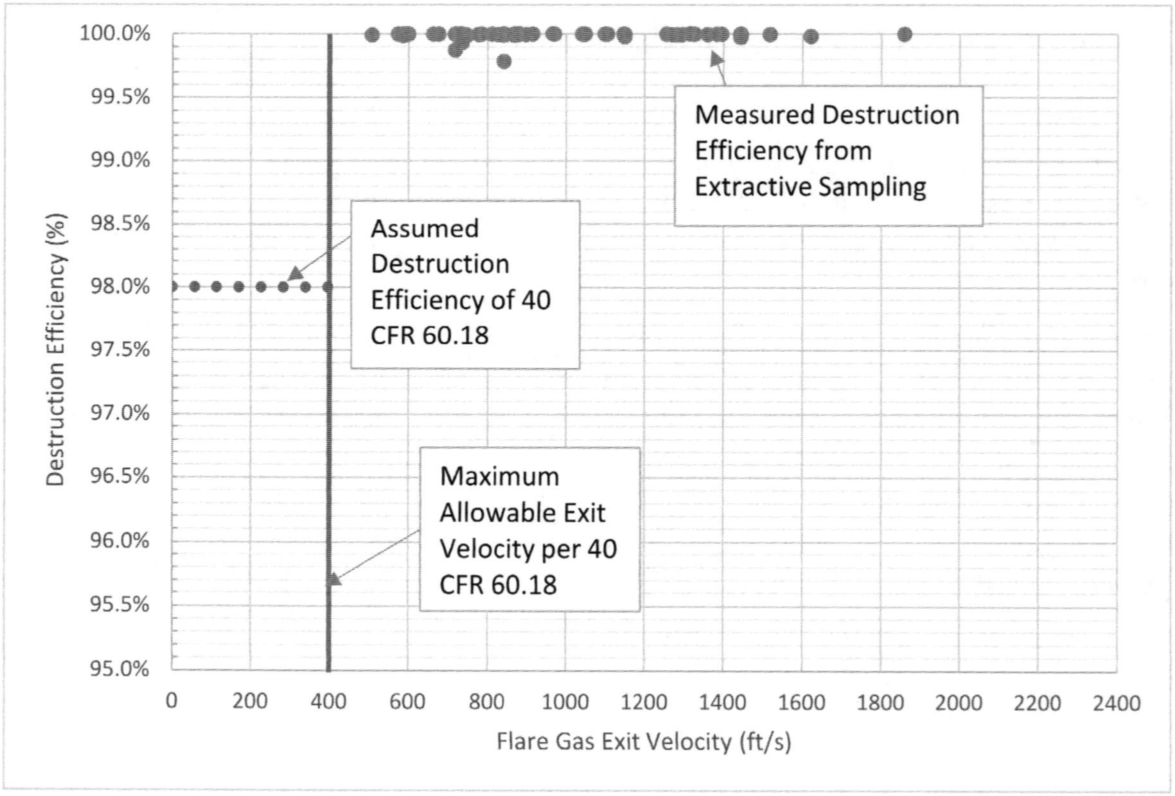

Figure 6 – Destruction and Removal Efficiency vs. Flare Gas Exit Velocity

The information in Figure 6 represents a total of 64 test runs. Flare gas mixtures ranged from molecular weight (MW) of 6.58 to 44.1 and from NHVs 440 to 2316 BTU/SCF. The mixtures were tested at operating pressures from 3 to 30 psig. The total compilation of all data was minimized by omitting the bottom 6% of the DRE values. The omitted test points will be discussed later.

Every gas combination tested that maintained a stable flame operated with high efficiencies despite the flare gas exit velocity exceeding the limits set forth in 40 CFR 60.18. The results from these tests on Zeeco's pressure-assisted tips prove that pressure-assisted flares should not have these exit velocity restrictions.

Several tests were performed to investigate the stability and effect of hydrogen (H_2) in hydrocarbon gas mixtures. The NHVs of the mixtures ranged from 440 to 1076 BTU/SCF with hydrogen mole percents between 25% and 70%. All gas mixtures were tested at exit velocities between 0.6 Mach (777 ft/s) and 1.0 Mach (2408 ft/s). Throughout the testing with H_2, the flame remained stable and attached to the tip and resulted in DRE and CE values of 99% or higher.

According to 40-CFR 60.18, an alternate method of calculating maximum exit velocity can be used for flare gas with a H_2 volume percent of 8.0 or greater. This alternate method linearly varies maximum exit velocity based on H_2 content between 8.0 percent at 25.6 ft/s and 15.6 percent at 122 ft/s and limits all volume percents greater than 15.6 to 122 ft/s. All of the H_2 mixtures tested had greater than 15.6 volume percent of H_2 and per 40 CFR 60.18 would be required to have an exit velocity less than 122 ft/s to meet the current EPA regulations. The results from Zeeco testing show that exit velocities greater than 122 ft/s did not negatively impact the flare efficiency for the H_2 mixtures tested. Although adding H_2 typically lowers the NHV of a gas mixture, testing showed that the presence of H_2 improved the stability of the flame compared to similar NHV mixtures without H_2. The benefit of H_2 flammability outweighed the downside of the reduced NHV, yielding high efficiencies.

The only unstable flame observed was when burning a gas mixture of propylene and nitrogen with an NHV of 600 BTU/SCF. This gas case was unable to maintain stability at the pressures tested and exhibited substantially lower DRE and CE than all other cases. In comparison, a mixture of natural gas and nitrogen with a NHV of 600 BTU/SCF remained stable at the pressures tested and maintained DRE and CE values above 99%. This suggests efficiency is not solely dependent upon the NHV of the flare gas. The difference in efficiency could be attributed to the volumetric ratio of flammable gas versus inert gas. In order to bring the NHV of a propylene mixture down to 600 BTU/SCF, a higher volume of inert gas is needed compared to a natural gas mixture. For example, the propylene mixture had approximately 73% volume inert gas where the natural gas mixtures had approximately 35% volume of inert gas. The higher volume of inert gas could be the cause of the lower efficiency.

Additional testing performed by Zeeco included the use of Passive Fourier Transform Infrared (PFTIR) in parallel with extractive sampling to measure CE. Four tests of propane, propylene, natural gas, and a propane and nitrogen mixture were used for this comparison. A similar exercise was conducted for Providence Engineering's new technology, FlareSENTRY.

The results from FlareSENTRY were extremely close to the results from the extractive method. All of the methods showed all four gas compositions had high CE while operating at sonic exit velocities as shown in Figure 7.

Gases	C3H8	C3H8/N2	C3H6	NG
NHV (BTU/SCF)	2316	1251	2183	937
40 CFR Maximum Allowable (ft/s)	400	400	400	400
Exit Velocity (ft/s)	841.4	969.9	869.8	1443.5
Mach Number	1.00	1.00	1.00	1.00
Flare Operating Pressure (psig)	16.0	10.3	16.9	15
CE (%) from Extractive Sampling	99.99%	99.99%	99.96%	99.99%
CE (%) from PFTIR	99.60%	99.90%	99.60%	99.50%
DRE (%) from Extractive Sampling	99.99%	99.99%	99.99%	99.99%
DRE (%) from FlareSENTRY	99.80%	99.55%	99.90%	99.70%

Figure 7 – Comparison of PFTIR vs. Extractive Sampling Method

FlareSENTRY is an optical-based monitoring technology developed by Providence Photonics, LLC, which remotely measures the DRE of flares. The unique capabilities of this new technology allow for continuous and autonomous monitoring of DRE with no operator input required. FlareSENTRY can also monitor the pilot ignition and provide a smoke index that can be used to predict the presence of smoke in the combustion plume. This technology will provide flare operators with real-time data to ensure optimal DRE while maintaining smokeless performance day or night. An industrial interface allows for integration with Distributed Control Systems (DCS) to enable closed loop operations. FlareSENTRY is a revolutionary new product that will redefine emissions monitoring and measurement for flares.

Conclusion

Multi-Point Ground Flares deliver smokeless flaring over a wide range of flows, compositions, and pressures while having minimal impact on surrounding communities. The testing conducted at the Zeeco Combustion and Research Test Facility investigated the effect of NHV and exit velocity on the stability and efficiency of Zeeco's pressure-assisted flare tips. All test results with stable flames demonstrate Zeeco's pressure-assisted flare tips provide high destruction and removal efficiency even at high exit velocities exceeding current regulatory limits. The extractive sampling results validated the performance of the new FlareSENTRY which is pioneering a practical method of monitoring and measuring flare efficiency. This testing has reinforced the industry consensus that pressure-assisted flare tips are capable of providing smokeless flaring with high destruction and removal and combustion efficiencies over a wide range of gas compositions and pressures. Recommendations for future testing are to investigate the stability of lower NHVs over a range of pressures and exit velocities as well as the effect of different ratios of combustible to inert gases. Current and future regulations should consider all available test results and allow the use of pressure-assisted flares with high exit velocities, up to and including Mach 1.0, without the need to perform an AMEL.

401845 Chloride Management in Naphtha Reforming with Continuous Catalyst Regeneration: A Method to Reduce Chloride Emissions and Consumption

Abstract:

Naphtha reforming with continuous catalyst regeneration (CCR) is an important process for the production of high octane gasoline and aromatics. Naphtha reforming catalysts are bi-functional catalysts that contain both acid and metal functions. The acid function is provided by chloride on the alumina base; as a result, chlorine containing compounds are present in the regenerator vent gas. In order to comply with environmental regulations many refiners must treat the regenerator vent gas to remove chlorine containing compounds. One method for reducing chloride emissions is the use of a conventional vent gas caustic scrubbing system which consumes caustic and generates hazardous waste. This contribution will compare a conventional vent gas scrubbing system to a system that captures chloride from the vent gas and recycles it back into the process, resulting in a 70% decrease in chloride consumption while simultaneously eliminating caustic usage.

399614 Electroreduction of Carbon Dioxide to Hydrocarbons Using Copper Nanoparticles' Electrode

Abstract:

A selective electro-reduction process of carbon dioxide in a three-phase system of gas-liquid-solid using a copper nanoparticle electrode is reported. This electrode was prepared by the surfactant-assisted hydrothermal reduction process with continuous ultra sonication. Prior to the electrochemical reduction of carbon dioxide, the gas was first absorbed in an organic-aqueous phase followed by adsorption on the surface of the copper nano electrode. The large surface area of the electrode and the pH of the solution both affect the electroreduction of carbon dioxide. The main products obtained were formic acid, ethylene, butane, methane and hydrogen with Faradic efficiencies of 27%, 10.5%, 0.3% and 36%, respectively. These faradic efficiencies were found to increase with an increase of: the organic to aqueous volume ratio, the solution temperature, the concentration of the electrolyte and the potential applied. The maximum conversion achieved was 22%.

Aqueous Ethylaminoethanol Blended Solvents: Promising Candidates for CO_2 Capture

(Principal Author)
Rahul Bhosale
Department of Chemical Engineering
College of Engineering
Qatar University, Doha, Qatar
rahul.bhosale@qu.edu.qa

(Other Authors)
Anand Kumar
Department of Chemical Engineering
College of Engineering
Qatar University, Doha, Qatar
akumar@qu.edu.qa

Fares A. AlMomani
Department of Chemical Engineering
College of Engineering
Qatar University, Doha, Qatar
falmomani@qu.edu.qa

Dareen Dardor
Department of Chemical Engineering
College of Engineering
Qatar University, Doha, Qatar
dd1202223@student.qu.edu.qa

Shahd Gharbia
Department of Chemical Engineering
College of Engineering
Qatar University, Doha, Qatar
sg1000354@student.qu.edu.qa

Moustafa Hussein Ali
Department of Chemical Engineering
College of Engineering

Qatar University, Doha, Qatar
ma1203561@student.qu.edu.qa

Jamila Folady
Department of Chemical Engineering
College of Engineering
Qatar University, Doha, Qatar
jf1003686@student.qu.edu.qa

Shiva Yousefi
Department of Chemical Engineering
College of Engineering
Qatar University, Doha, Qatar
sy1106491@student.qu.edu.qa

Mehak Jilani
Department of Chemical Engineering
College of Engineering
Qatar University, Doha, Qatar
mj1104333@student.qu.edu.qa

Prepared for Presentation at
American Institute of Chemical Engineers
2015 Spring Meeting
11th Global Congress on Process Safety
Austin, Texas
April 27-30, 2015

UNPUBLISHED

Aqueous Ethylaminoethanol Blended Solvents: Promising Candidates for CO_2 Capture

Rahul Bhosale
Department of Chemical Engineering
College of Engineering
Qatar University, Doha, Qatar
rahul.bhosale@qu.edu.qa

Anand Kumar, Fares A. AlMomani, Dareen Dardor, Shahd Gharbia, Moustafa Hussein Ali, Jamila Folady, Shiva Yousefi, Mehak Jilani

Department of Chemical Engineering
College of Engineering
Qatar University, Doha, Qatar

Keywords: Ethylaminoethanol, CO_2 Capture, Fall in Pressure Technique, Reactive Absorption

Abstract

This paper reports the investigation of effect of temperature on physico-chemical properties of formulated solvent (comprises on ethylaminoethanol, N-methyl-2-pyrollidone, and $K_2CO_3/KHCO_3$) and reaction kinetics of the absorption of CO_2 in this formulated solvent. The physico-chemical properties were estimated by using a specific gravity bottle (density), an Ostwald's capillary viscometer (viscosity), and a stirred cell reactor (diffusivity and solubility of CO_2 in formulated solvent). Similarly, a stirred cell reactor was used for the investigation of CO_2 absorption kinetics by following a novel fall in CO_2 pressure technique.

1. Introduction

Due to the increasing atmospheric concentration of CO_2 and its adverse effect on the environment, significant efforts are underway towards minimization of CO_2 release [1]. Among the various techniques investigated towards reduction of CO_2 emission, absorption of CO_2 using amine solutions is industrially accepted and shows very less complications as compared to other methods such as physical adsorption [2-3] and membrane separation [4].

Among all previously studied amines [5-9], MEA remains the most popular for CO_2 capture owing to its higher reactivity and lower solvent cost as compared to that of the other amines. However, it also possesses high vapor pressure which makes it unsuitable for low pressure operations due to the likely vaporization losses. In addition, due to the possibility of the reaction with the reactor vessels, tubings, and several process compartments, utilization of highly concentrated MEA solutions for acid gas sweetening is not feasible. The inferior CO_2 loading capacity, elevated solvent regeneration cost, formation of toxic products due to the degradation

Note: Do not add page numbers. Do not refer to page numbers when referencing different portions of the paper

in O_2-rich environment (especially in case of flue gas), reduced scrubber efficiency due to amine decomposition and corrosion in the equipments and piping's are few of the other limitations related to MEA, which induces the quest for alternative absorbents.

In our recent investigations, we realized that aqueous solution of ethylaminoethanol (EAE) is a very potential solvent for the capture of CO_2. It shows lower degradation in presence of CO_2 and shows equivalent CO_2 absorption capacity as compared to the aqueous MEA [10-11]. In this paper, we attempted to improve the CO_2 absorption capacity of EAE by formulating a solvent containing aqueous mixture of EAE, N-methyl-2-pyrollidone (NMP), and potassium salts of carbonate ($K_2CO_3/KHCO_3$). The physico-chemical properties and CO_2 absorption capacity of this formulated solvent was investigated at different temperatures using a stirred cell reactor..

2. Materials & Methods

2.1 Materials

EAE (99% pure), anhydrous K_2CO_3 (analytical reagent grade) and $KHCO_3$ (laboratory grade) were procured. Nitrous oxide (N_2O), N_2, O_2, and CO_2 with minimum stated purities equal to 99.5% were obtained and used during degradation and kinetics experiments. All the chemicals were used without any pretreatment and all the solutions were prepared using distilled water.

2.2 Experimental Set-Up for Kinetic Study and Estimation of Physico-chemical Properties

The experiments pertaining to the kinetics of the absorption of CO_2 in aqueous formulated solvent containing EAE, NMP, and $K_2CO_3/KHCO_3$ were carried out in a stirred cell reactor (diameter 97 mm, height 187 mm) as shown in Figure 1. These experiments were carried out by operating the reactor in a batch mode with flat horizontal gas-liquid interface. The total volume of the reactor was 1450 ml and the interfacial surface area was 7.4×10^{-3} m^2. The reactor was equipped with a flange made up of SS 316 and fabricated by Sharad Autoclave Engineers, Mumbai, India. A zero leak variable speed magnetic drive (Switzerland) was mounted on this flange. Provisions were made for the gas and liquid inlet and outlet ports and the entire stirred cell assembly was tested to be leak-proof. A pressure transducer (Trans Instruments, UK, 0-1 bar) was mounted on the flange and coupled with a data acquisition system which enabled the measurement of the fall in CO_2 pressure during the absorption reaction. The uncertainty in the pressure measurement was around ± 1 mbar. The reactor was stirred with two disc blade turbine impellers (six blades, diameter 41 mm) mounted on the same shaft, one each for the gas phase and the liquid phase. The stirrer was driven by 1/20 HP geared motor with a maximum stirring speed of 165 rpm. The speed of agitation could be adjusted to the desired value with an accuracy of ± 1 rpm. The impeller speed during kinetic measurements was limited to a certain rpm so as to ensure that the gas-liquid interface was undisturbed, which further results into a geometrically simple and hence exactly known interfacial area. During the solubility measurements, a Remi Motor with a maximum speed of agitation of 3000 rpm was used. The stirred cell assembly was immersed in a constant temperature water bath to ensure isothermal conditions and the temperature was controlled to within ± 0.1°C of the set point by means of a temperature indicator and controller (TIC). A gas feed coil was immersed in the water bath to enable the solute gas to attain the reactor temperature prior to being charged inside the reactor.

2.3 Experimental Procedure for Solubility Measurement

All the experiments associated with the measurement of solubility of N_2O in an aqueous formulated solvent were conducted using the same experimental set-up shown in Figure 1. At first, 250 ml of the absorbing solvent was loaded in the reactor and it allowed attaining the desired reaction temperature. N_2O at a chosen partial pressure was charged inside the reactor and the initial pressure within the stirred cell was recorded with the help of pressure transducer and the data acquisition system. The reactor contents were then agitated at higher speeds of 1200 rpm to ensure an attainment of the equilibrium. The equilibrium was assumed to be reached when the digital pressure indicator showed a constant reading of the system pressure for 3 h. This was also confirmed by the 'P_{CO2} vs. t' data recorded by the data acquisition system. This technique was validated by performing solubility measurements of CO_2 and N_2O in water and comparing the results with the published data. The error between the H_{CO2} in water determined during this investigation and the previously reported H_{CO2} in water was ± 3%, which is well within the acceptable limit. The reproducibility of the results was also examined and the error in all experimental measurements was observed to be less than ± 2%.

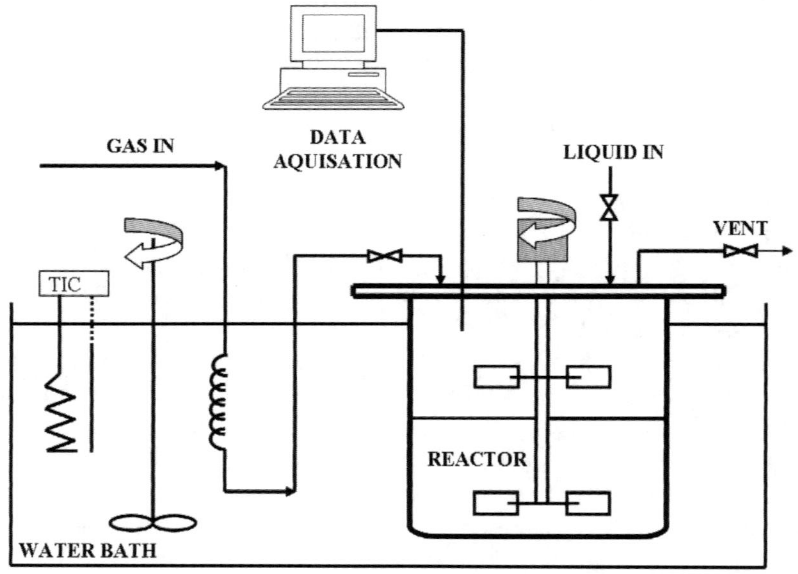

Figure 1. Experimental set-up for solubility and kinetics study.

2.4 Experimental Procedure for Kinetics Measurements

A series of experiments were performed to study the kinetics of the absorption of CO_2 in an aqueous formulated solvent. The reactor was first charged with 250 ml of the absorbing solvent and the reactor contents were allowed to attain the desired absorption temperature. CO_2 from the gas cylinder was then loaded inside the stirred cell reactor at preferred partial pressure which was recorded with the help of the data acquisition system and this step was considered to be the starting point of the reaction. The reactor content was then stirred at a pre-determined speed of agitation while ensuring that the gas-liquid interface was not disturbed. The decrease in the system pressure due to reactive absorption of CO_2 was monitored with the help of pressure transducer and 'P_{CO2} vs. t' data was recorded for 30 seconds using the data acquisition system. The 'P_{CO2} vs. t' data from t = 5 seconds to t = 25 seconds were plotted and fitted to a third degree polynomial using the least squares regression and the rate of absorption of CO_2 was calculated with the help of the slope ($-dP_{CO2} / dt$). This measurement, which is thoroughly based on the fall in pressure of CO_2, enables a simple and straight forward estimation of the absorption rate. Further analysis of the liquid phase is not required and the decrease in the system pressure is the only factor considered for the evaluation of kinetic parameters. The reproducibility of the results was examined and the error in all experimental measurements was less than ± 3%.

3. Results & Discussion

3.1 Estimation of Physicochemical Properties

To determine kinetic parameters, estimation of physico-chemical properties such as density (ρ) and viscosity (μ) of the formulated solvent and solubility (H_{CO2}) and diffusivity (D_{CO2}) of CO_2 in formulated solvent is necessary. Initially, ρ and μ of aqueous EAE and formulated solvent was measured by using a specific gravity bottle and an Ostwald's capillary viscometer. The effect of EAE concentration (0.5 to 2 kmol/m^3) on ρ and μ of formulated solvent was investigated at 308 K and the results obtained are shown in Figure 2. Similarly, the variation in the ρ and μ of solvents at different temperatures (from 298 to 323 K) with EAE concentration = 1.5 kmol/m^3 is presented in Figure 3. According to the results reported, it was observed that the density of formulated solvent decreases with the increase in the EAE concentration and temperature. Viscosity of the formulated solvent increases with the increase in the EAE concentration and decreases with respect to temperature.

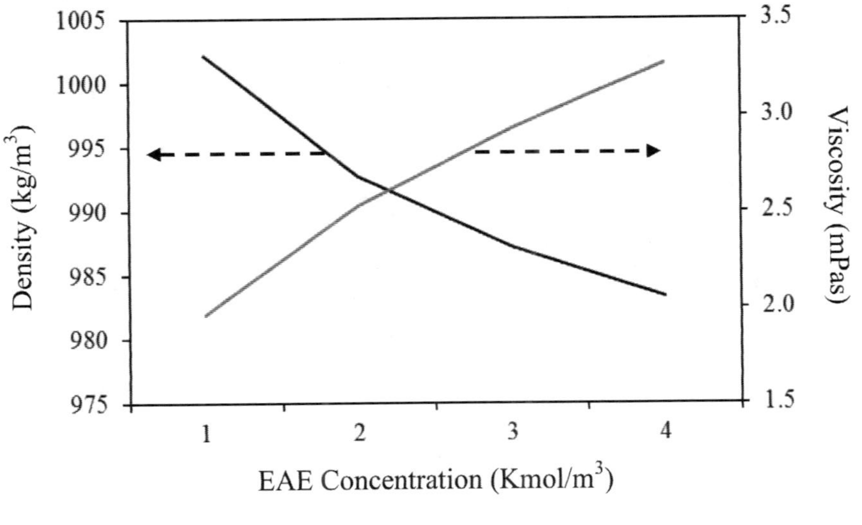

Figure 2. Density and viscosity of formulated solvent (Effect of EAE concentration).

The D_{CO2} in aqueous EAE and formulated solvent was determined using the N_2O analogy. The D_{CO2} and D_{N2O} in water was estimated using the following expressions:

$$D_{CO_2} = (2.35\times10^{-6})\times\exp\left(\frac{-2119}{T}\right) \quad (1)$$

$$D_{N_2O} = (5.07\times10^{-6})\times\exp\left(\frac{-2371}{T}\right) \quad (2)$$

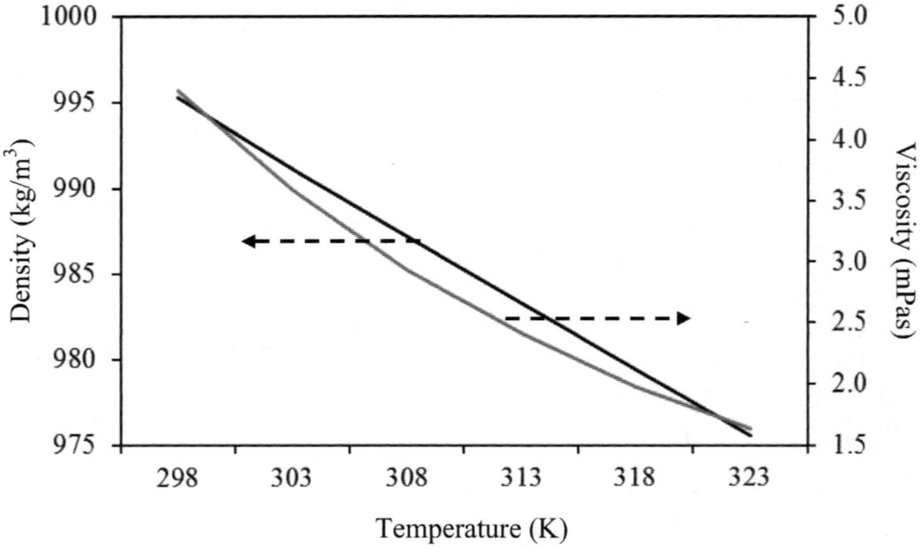

Figure 3. Density and viscosity of formulated solvent (Effect of Temperature).

By using the results obtained via viscosity measurement, the D_{CO2} and D_{N2O} in aqueous EAE and formulated solvent was estimated using modified Stokes-Einstein correlation:

$$\left(D_{N_2O}\times\mu^{0.8}\right)_{amine} = \left(D_{N_2O}\times\mu^{0.8}\right)_{H_2O} \quad (3)$$

$$\left(D_{CO_2}\times\mu^{0.8}\right)_{amine} = \left(D_{CO_2}\times\mu^{0.8}\right)_{H_2O} \quad (4)$$

The influence of temperature on D_{CO2} in formulated solvent was experimentally studied and the results obtained are shown in Figure 4. As the temperature increases from 298 K to 323 K, the D_{CO2} increases by 61.5%.

The N_2O analogy was also used for the estimation of H_{CO2} in aqueous formulated solvent. Firstly, the H_{N2O} in aqueous formulated solvent was assessed by performing the solubility experiments in the stirred cell reactor. The initial and final readings of pressure of N_2O within the reactor were recorded using the data acquisition system and no alteration in the total pressure of the stirred cell over an extended period is deemed as the attainment of the equilibrium. Further, the H_{CO2} can be computed according to the following correlation:

$$H_{CO2} = \gamma \times H_{N2O} \tag{5}$$

The solubility measurements were validated by comparing the experimentally obtained H_{N2O} and H_{CO2} in water at 303 K with the published literature. To begin with, the H_{CO2} in water was estimated experimentally at 303 K and it matches very well with the H_{CO2} in water reported by Versteeg and van Swaalj [12].

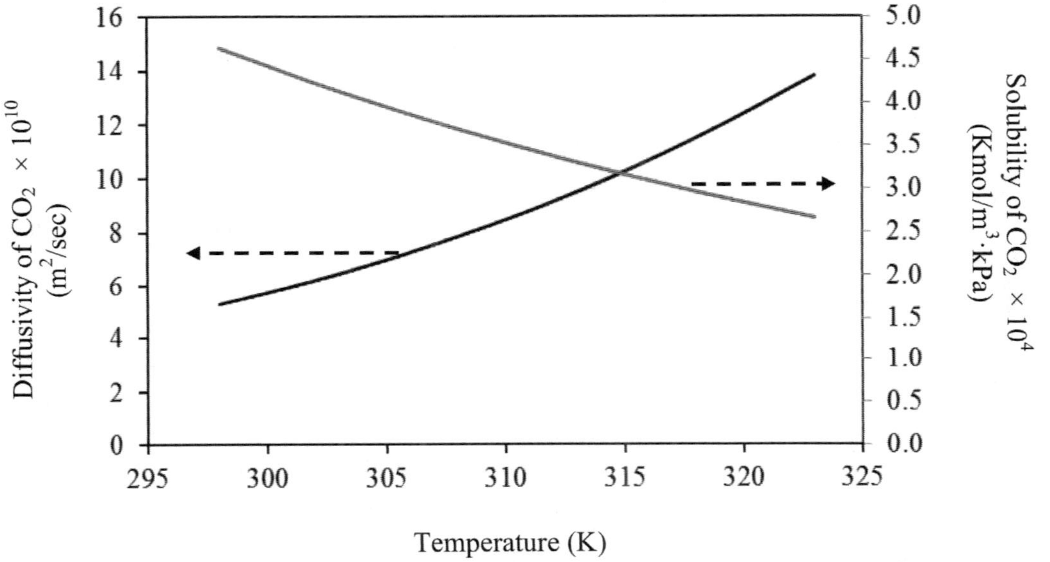

Figure 4. Diffusivity and Solubility of CO_2 in Formulated Solvent (Effect of Temperature).

Likewise, the experiments were performed to determine H_{N2O} in aqueous formulated solvent and via N_2O analogy, the H_{CO2} was inferred. Figure 4 shows the alteration in H_{CO2} in formulated solvent with respect to temperature. The results presented indicate increase in the temperature results into decrease in the H_{CO2} in aqueous formulated solvent. Furthermore, the liquid side mass transfer coefficient for the CO_2 – aqueous formulated solvent reaction system was determined using the techniques earlier adopted [13-14] and observed to be equal to 2.02×10^{-4} m/s.

3.2 Investigation of Absorption Kinetics

A stirred cell reactor and fall in pressure technique was used for the investigation of the reaction kinetics of the reactive absorption of CO_2 in formulated solvent. Table 1 lists the experimental parameters used during this study. At first the effect of speed of agitation on absorption kinetics related to CO_2-formlated solvent was examine. Results obtained during this study indicate that reactions associated with the CO_2-formulated solvent system belong to the fast reaction regime. Also, the effect of speed of agitation (from 70 – 110 rpm) was insignificant and the rate of CO_2 absorption remained unchanged. Therefore, all the experiments related to the investigation of the effect of temperature on absorption of CO_2 using a formulated solvent were performed at speed of agitation equal to 80.

Table 1. Experimental parameters for the CO_2 absorption experiments.

Experimental parameters	Values
EAE concentration	2 $kmol/m^3$
NMP concentration	5 $kmol/m^3$
$K_2CO_3/KHCO_3$	10wt% of EAE
Temperature	303 to 318 K
Partial pressure of CO_2	4 kPa
Agitation speed	80 rpm

Influence of reaction temperature on rate of absorption of CO_2 in formulated solvent was investigated experimentally and the findings are presented in Figure 5. According to the obtained results, as the reaction temperature increased from 303 K to 318 K, the rate of absorption of CO_2 in formulated solvent is also increased by a factor of 1.734.

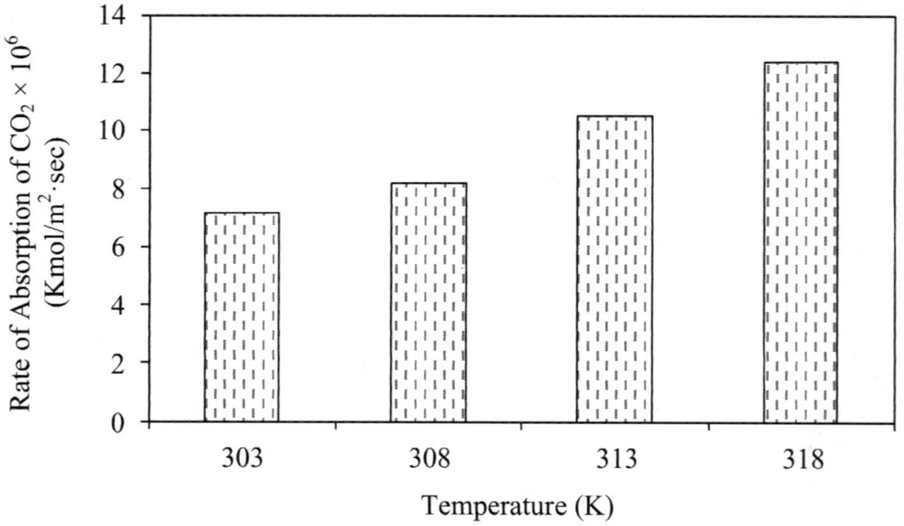

Figure 5. Effect of temperature on rate of absorption of CO_2 in formulated solvent.

4. Conclusion

In this investigation, the reaction kinetics of absorption of CO_2 in formulated solvent was studied. A glass stirred cell reactor and fall in pressure technique was used and effect of temperature on CO_2 absorption rate was investigated in detail. At first, the physicochemical properties of formulated solvent (density, viscosity, diffusivity and solubility of CO_2) were estimated. Following effects were realized due to the increase in the temperature: density decreases, viscosity decreases, solubility of CO_2 decreases, and diffusivity of CO_2 increases. Similarly, with the increase in the reaction temperature, the rate of absorption of CO_2 in formulated solvent increased by 1.734.

5. Acknowledgments

The authors gratefully acknowledge the financial support provided by the Qatar University Internal Grants QUUG-CENG-CHE-13/14-4 and QUUG-CENG-CHE-14\15-10.

References

[1] Strazisar, B. R., Anderson, R. R., White, C. M., "Degradation pathways for monoethanolamine in a CO_2 capture facility", *Energy & Fuels*, Volume 17, Page No. 1034, 2003.

[2] Wurzbacher, J.A., Gebald, C., Steinfeld, A., "Separation of CO_2 from air by temperature vacuum swing adsorption using diaminefunctionalized silica gel", *Energy & Environmental Science*, Volume 4, Page No. 3584, 2011.

[3] Lively, R.P., Chance, R.R., Kelley, B.T., Deckman, H.W., Drese, J.H., Jones, W.C., Koros, W.J., "Hollow fiber adsorbents for CO_2 removal from flue gas", *Industrial & Engineering Chemistry Research*, Volume 48, Page No. 7314, 2009.

[4] Hagg, M., Lindbrathen, A., "CO_2 capture from natural gas fired power plants by using membrane technology", *Industrial & Engineering Chemistry Research*, Volume 44, Page No. 7668, 2005.

[5] Vaidya, P. D., Mahajani, V. V., "Kinetics of the reaction of CO_2 with aqueous formulated solution containing monoethanolamine, N-methyl-2-pyrrolidone, and diethylene glycol," *Industrial & Engineering Chemistry Research*, Volume 44, Page No. 1868, 2005.

[6] Al-Juaied, M., Rochelle, G. T., "Absorption of CO_2 in aqueous diglycolamine. *Industrial & Engineering Chemistry Research*, Volume 45, Page No. 2473, 2005.

[7] Rowley, R. L., Adams, M. E., Marshall, T. L., Oscarson, J. L., Wilding, W. V., Anderson, D. J., "Measurement of the absorption rate of carbon dioxide into aqueous diethanolamine," *Journal of Chemical Engineering Data*, Volume 43, Page No. 427, 1998.

[8] Horng, S., Li, M., "Kinetics of absorption of carbon dioxide into aqueous solutions of monoethnolamine + triethanolamine," *Industrial & Engineering Chemistry Research*, Volume 41, Page No. 257, 2002.

[9] Mandal, B., Bandyopadhyay, S. S., "Simultaneous absorption of CO_2 and H_2S into aqueous blends of N-methyldiethanolamine and diethanolamine," *Environmental Science & Technology*, Volume 40, Page No. 6076, 2006.

[10] Bhosale, R., Mahajani, V. V., "Kinetics of absorption of carbon dioxide in aqueous solution of ethylaminoethanol modified with N-methyl-2-pyrolidone", *Separation Science & Technology*, Volume 48, Page No. 2324, 2013.

[11] Bhosale, R., Mahajani, V. V., "Kinetics of thermal degradation of renewably prepared amines useful for flue gas treatment", *Journal of Renewable & Sustainable Energy*, Volume 5, Page No. 063110-1, 2013.

[12] Versteeg, G. F., van Swaaij, W. P. M., "Solubility and diffusivity of acid gases (carbon dioxide, nitrous oxide) in aqueous alkanolamine solutions," *Journal of Chemical Engineering Data*, Volume 33, Page No. 29, 1988.

[13] Vaidya, P. D., Kenig, E. Y., "Absorption of CO_2 into aqueous blends of alkanolamines prepared from renewable resources," *Chemical Engineering Science*, Volume 62, Page No. 7344, 2007.

[14] Vaidya, P. D., Mahajani, V. V., "Kinetics of the reaction of CO_2 with aqueous formulated solution containing monoethanolamine, N-methyl-2-pyrrolidone, and diethylene glycol," Industrial & Engineering Chemistry Research, Volume 44, Page No. 1868, 2005.

401212 Solar Steam Reforming for Hydrogen Production Using Solar Salts As Heat Transfer Fluid

Abstract:

Introduction

The application of solar-thermal power to directly drive heat demanding thermochemical conversion is one of the most rational ways to exploit solar energy. This approach enhances the reduction of carbon footprint of chemical conversion, and allows the chemical storage of solar energy. Particularly, in fuel refinery it is possible to improve the heating value and the environmental impact (i.e. the overall "quality") of the primary feedstock by the aid of solar energy: in this case the final product, often called "solar fuel", will partially or totally drive solar energy in its chemical energy.

Additionally, the worldwide hydrogen demand is increasing, and its use in Fuel Cells (FCs) is expected to make hydrogen one of the foremost fuels in a future more sustainable energy system, mainly thanks to its low environmental impact when used in FC systems.

On the other hand, the extensive introduction of hydrogen as energy vector is limited by the actual lack (and costs) of a reliable hydrogen distribution infrastructure. This barrier can be surmounted by the development of systems for decentralized hydrogen production (i.e. close to the end-user). Fuel-flexibility and the possibility to power the process with renewable energy sources (solar and/or biomass) are two additional preconditions for sustainable hydrogen production.

The above considerations are on the basis of the project CoMETHy (Compact Multifuel-Energy to Hydrogen converter) co-funded by the European Commission under the Fuel Cells and Hydrogen Joint Undertaking (FCH JU). Indeed, CoMETHy's general objective is to support the intensification of hydrogen production processes, by developing an innovative compact solar steam reformer to convert reformable fuels (methane, ethanol, etc.) to pure hydrogen, according to the following general reaction scheme:

$CnH_{2n+2} + 2nH_2O$ à $nCO_2 + (3n+1)H_2$

where C_nH_{2n+2} is the hydrocarbon feedstock that can be even replaced by an oxygenated chemical like ethanol or glycerol.

In case of unavailability of the solar radiation, the reformer should be adaptable to other back up fuels (e.g. natural gas, biomass), depending on the locally available energy mix, in order to keep the desired hydrogen production rate.

Process concept and results

In principle, the process under development is based on low-T steam reforming. Process heat is supplied by means of a low-cost and environmentally friendly liquid heat transfer fluid, i.e. the binary Molten Salts (MS) mixture $NaNO_3/KNO_3$ (60/40 w/w), often referred as "solar salt". This allows coupling with Concentrating Solar Power (CSP) plants using this molten salts mixture as thermal storage system at temperatures up to 550°C. Accordingly, this fluid collects heat from the different available heat sources and transfers it to the thermochemical plant. A suitable heat storage system, based on the use of the solar salt mixture, allows mismatch between the fluctuating solar source and the often steady running chemical plant: in principle, this makes possible to drive the thermochemical plant at steady state, regardless of the effective instantaneous availability of the solar radiation, even overnight and during cloudy periods of time. Clearly, for a given location of the plant (i.e. given yearly solar radiation characteristics), there is an interplay between the nominal power of the CSP plant (i.e. the solar field area), the heat storage capacity and the amount of back up fuel to balance the solar radiation lacks: this is on the bases of CSP plant hybridization, which can be either achieved by a biomass (or RDF, Refuse Derived Fuel) or fossil fuel (e.g. off-gas) combustor and a best compromise between the CSP and the back-up fuel duties should be identified for the technical-economical optimization of the process [1,2]. Selective membranes allow recovery of high-grade hydrogen and increase conversion despite the relatively low operating temperatures.

If compared to a typical steam reforming process, this steam reforming technology operates at lower temperatures, from typical 850-950°C down to 400-550°C, with consequent significant gain in material costs since no special steel alloys for high-temperature operation are required. The high-temperature furnace is then replaced by a flameless heat exchanger heated by a liquid molten salts stream, making the whole reactor envelope more compact. Additionally, by operating at lower temperatures, it is possible to combine steam reforming and water-gas-shift (WGS) reactions into a single stage at 400-550°C, resulting into a low outlet CO content (<10 %vol) and a reduction of the reformer heat duty. The integration with membranes avoids dedicated hydrogen separation and purification units, to further improve the compactness and enhance the conversion despite the thermodynamic limitations of low-temperature reforming.

When the thermochemical plant is powered by solar heat, in principle, there will be no combustion in the whole process, and no combustion generated CO_2-containing flue gases emitted to the atmosphere: this will result in reduction of fuel consumption and greenhouse gas (GHG) emissions being in the order of 40% to more than 50% compared to the conventional route [3].

The reduction of fossil fuel consumption will make the hydrogen production cost less sensitive to the fossil (e.g. NG) price in the solar steam reforming: therefore a breakeven point for the hydrogen production cost is foreseen, resulting in a solar route economically more convenient than the fossil based one [4]. The same solar steam reforming process applied to biomass derived fuels (i.e. bio-fuels like biogas, bioethanol, etc.) allows totally renewable hydrogen production.

An additional advantage of the use of molten salts as heat transfer fluid for steam reforming is when some stand-by periods of the plant are foreseen, as it is the case of small-medium scale reformers for hydrogen refueling stations. In this case, the continuous molten salts recirculation eases the overall process management by maintaining all plant components (e.g. catalyst, membrane) at working temperature (400-550°C) also during stand-by periods using: this minimized start-up periods and ageing due to thermal cycling.

The development of this technology, however, involves a number of interconnected R&D challenges to be faced. Three reference feedstocks have been chosen in the perspective of the multi-fuel application: methane-rich gas (e.g. natural gas), CH_4/CO_2 mixtures (up to 50%vol. CO_2, representing biogas), and bioethanol (i.e. diluted ethanol). First, it is necessary to identify and develop suitable catalysts and membranes for this application. Second, the catalyst and the membrane will be integrated in the design of a molten salts heated steam reformer. Finally, the best strategies to couple this reformer with a CSP plant will be studied.

Extensive catalyst studies led to the identification of promising catalyst materials suitable for the low-temperature steam reforming of methane (biogas) and ethanol. Open foam ceramic silicon carbide foams have been successfully developed and characterized as catalyst supports to enhance the heat transfer and to minimize pressure drops. In parallel, relevant effort has been addressed to the development of low-cost and suitable hydrogen selective membranes for CoMETHy application, basically composite membranes consisting of a thin Pd (alloy) layer (2-5 μm) coated on a porous ceramic or stainless steel support. Different membrane reactor configurations have been conceived and the performances evaluated experimentally.

Summary and conclusions

An innovative solar reformer is being developed in the CoMETHy project. Besides the interesting potentials, its development involves different research topics: selection of advanced catalyst systems with enhanced heat transfer capability and selective composite membranes, and the identification of the best options for catalyst/membrane assembly, and reformer coupling with the CSP plants.

Acknowledgements

CoMETHy project has received funding from the European Union's Seventh Framework Programme (FP7/2007-2013) for the Fuel Cells and Hydrogen Joint Technology Initiative under grant agreement n. 279075.

References

[1] Giaconia A, Grena R, Lanchi M, Liberatore R, Tarquini P. Hydrogen/methanol production by sulfur–iodine thermochemical cycle powered by combined solar/fossil energy, Int J Hydrogen Energy 2007;32:469-81.

[2] De Falco M., Giaconia A., Marrelli L., Tarquini P., Grena R., Caputo G. Enriched methane production using solar energy: an assessment of plant performance. Int J Hydrogen Energy 2009;34:98-109.

[3] Giaconia A., De Falco M., Caputo G., Grena R., Tarquini P., Marrelli L. Solar steam reforming of natural gas for hydrogen production using molten salt heat carriers. AIChE J 2008;54:1932-44.

[4] Moeller S, Kaucic D, Sattler C. Hydrogen production by solar reforming of natural gas: a comparison study of two possible process configurations. ASME J Sol Energy Eng 2006;128:16–23.

15 AIChE
Spring Meeting, Austin, TX

Solar Hydrogen Production via Thermochemical Metal Oxide – Metal Sulfate Water Splitting Cycle

(Principal Author)
Rahul Bhosale
Department of Chemical Engineering
College of Engineering
Qatar University, Doha, Qatar
rahul.bhosale@qu.edu.qa

(Other Authors)
Anand Kumar
Department of Chemical Engineering
College of Engineering
Qatar University, Doha, Qatar
akumar@qu.edu.qa

Fares A. AlMomani
Department of Chemical Engineering
College of Engineering
Qatar University, Doha, Qatar
falmomani@qu.edu.qa

Dareen Dardor
Department of Chemical Engineering
College of Engineering
Qatar University, Doha, Qatar
dd1202223@student.qu.edu.qa

Shahd Gharbia
Department of Chemical Engineering
College of Engineering
Qatar University, Doha, Qatar
sg1000354@student.qu.edu.qa

Moustafa Hussein Ali
Department of Chemical Engineering
College of Engineering
Qatar University, Doha, Qatar
ma1203561@student.qu.edu.qa

Jamila Folady
Department of Chemical Engineering
College of Engineering
Qatar University, Doha, Qatar
jf1003686@student.qu.edu.qa

Shiva Yousefi
Department of Chemical Engineering
College of Engineering
Qatar University, Doha, Qatar
sy1106491@student.qu.edu.qa

Mehak Jilani
Department of Chemical Engineering
College of Engineering
Qatar University, Doha, Qatar
mj1104333@student.qu.edu.qa

Diana Abuarja
Department of Chemical Engineering
College of Engineering
Qatar University, Doha, Qatar
da1001802@student.qu.edu.qa

Prajakta Angre
Department of Chemical Engineering
MGM College of Engineering & Technology
Mumbai University, Navi Mumbai
prajaktaangre752007@gmail.com

Prepared for Presentation at
American Institute of Chemical Engineers
2015 Spring Meeting
11th Global Congress on Process Safety
Austin, Texas

April 27-30, 2015

UNPUBLISHED

Solar Hydrogen Production via Thermochemical Metal Oxide – Metal Sulfate Water Splitting Cycle

Rahul Bhosale[1]
Department of Chemical Engineering
College of Engineering
Qatar University, Doha, Qatar
rahul.bhosale@qu.edu.qa

Anand Kumar[1], **Fares A. AlMomani**[1], **Dareen Dardor**[1], **Shahd Gharbia**[1], **Moustafa Hussein Ali**[1], **Jamila Folady**[1], **Shiva Yousefi**[1], **Mehak Jilani**[1], **Diana Abuarja**[1], **Prajakta Angre**[2]

[1]Department of Chemical Engineering
College of Engineering
Qatar University, Doha, Qatar

[2]Department of Chemical Engineering
MGM College of Engineering & Technology
Mumbai University, Navi Mumbai, India.

Keywords: Solar Fuels, Iron Oxide – Iron Sulfate Cycle, Water Splitting, Computational Thermodynamic Modeling

Abstract

This paper reports the thermodynamic analysis of solar H_2 production via two-step thermochemical iron oxide – iron Sulfate (IO-IS) water splitting cycle. The first step belongs to the exothermic oxidation of FeO via SO_2 and H_2O producing FeSO4 and H_2 and second step corresponds to the endothermic reduction of FeSO4 into FeO, SO_2, and O_2. The products, FeO and SO_2 can be recycled to step 1 and hence, reutilized for the production of H_2 via water splitting reaction. Thermodynamic equilibrium compositions of the thermal reduction and water splitting reactions were computed as a function of reaction temperatures. Second law thermodynamic analysis was performed to determine the $_{absorption}$, Q_{solar}, $Q_{reactor-net}$, $Q_{re-radiation}$, $Q_{FeO-oxd}$, and η_{cycle} and $\eta_{solar-to-fuel}$ respectively.

1. Introduction

H_2 is considered as one of the potential energy carriers due to its capacity of producing 143 MJ/kg of energy (higher in comparison with the oil, gas, and coal individually) and its environmentally friendly nature (burning of H_2 produces water with no other polluting gases).

Note: Do not add page numbers. Do not refer to page numbers when referencing different portions of the paper

Water splitting reaction is considered as one of the most promising ways for the production of H_2. Several thermochemical cycles such as iron oxide cycle [1-6], zinc/zinc oxide cycle [7-10], tin/tin oxide cycle [8, 11-13], mixed ferrite cycle [14-21], ceria cycle [22-24], sulfur-iodine cycle [25-27], and hybrid sulfur cycle [28-29] were investigated towards H2 production via water-splitting reaction. Among these, the sulfur-iodine cycle (reaction set I) and its variation the hybrid sulfur cycle (reaction set II) are more appealing as the required operating temperatures are lower as compared to other thermochemical cycles.

Reaction set I: sulfur-iodine cycle

$H_2SO_4 \rightarrow SO_3 + H_2O$	(673K)	(1)
$SO_3 \leftrightarrow SO_2 + {}^1/_2\, O_2$	(1123 – 1273K)	(2)
$SO_2 + 2H_2O + I_2 \rightarrow H_2SO_4 + 2HI$	(393K)	(3)
$2HI \rightarrow H_2 + I_2$	(573 – 723K)	(4)

Reaction set II: hybrid sulfur cycle

$H_2SO_4 \rightarrow SO_3 + H_2O$	(673K)	(5)
$SO_3 \leftrightarrow SO_2 + {}^1/_2\, O_2$	(1123 – 1273K)	(6)
$SO_2 + 2H_2O \rightarrow H_2SO_4$	(353 – 393K)	(7)

This paper proposes the utilization of the 'iron oxide – iron sulfate' (IO-IS) cycle for the production of solar H_2 via thermochemical water splitting reaction. In this study, the thermodynamic feasibility of this cycle was investigated and obtained results are presented in detail. At first, the thermodynamic equilibrium compositions during the solar thermal reduction of $FeSO_4$ under inert atmosphere and oxidation of FeO via water splitting reaction were determined. Furthermore, the maximum theoretical solar energy conversion efficiency of the IO-IS cycle is determined by performing the second law thermodynamic analysis over different solar reactor temperatures and with/without considering the heat recuperation.

2. Chemical Thermodynamic Equilibrium

2.1 Thermal Reduction of Iron Sulfate

Figure 1 represents that solar thermal reduction of $FeSO_4$ into FeO, $SO_2(g)$ and $O_2(g)$ is possible at 1250K when 100 moles/sec of inert Ar flowrate is used. The thermal reduction reaction is represented as follows:

$$FeSO_4 \rightarrow FeO + SO_2(g) + {}^1/_2\, O_2(g) \qquad (1)$$

2.2 Water Splitting Reaction

Figure 3 represent variations in the equibrium composition for water splitting reaction (in absence of Ar). According to the results presented, the production of H_2 via oxidation of FeO by SO_2 and H_2O is possible below 483K. It is further observed that at lower temperatures, $FeSO_4$ and H_2 are the major species, whereas the molar concentration of FeO, H_2O and SO_2 increases with the increase in water splitting temperature.

$$FeO + SO_2(g) + H_2O(g) \rightarrow FeSO_4 + H_2(g) \qquad (2)$$

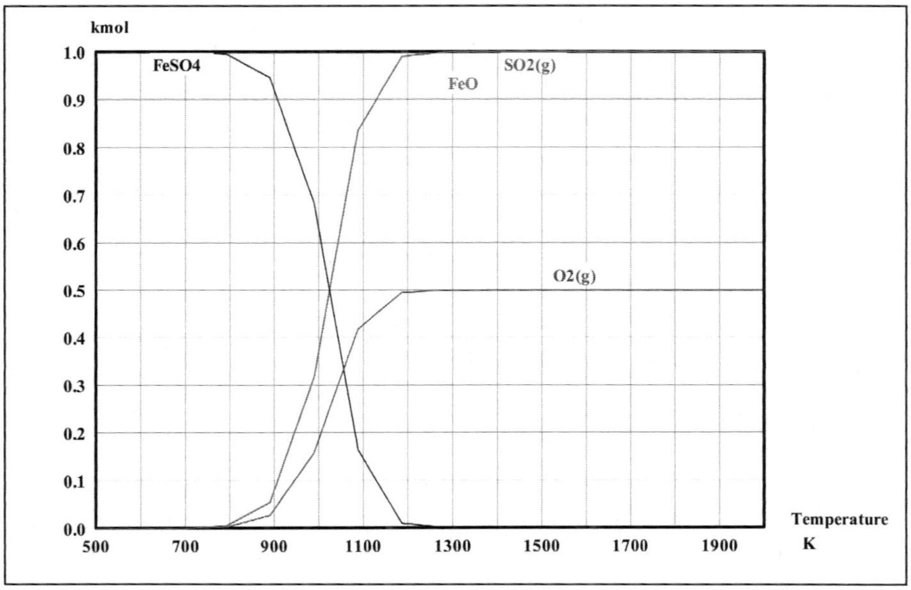

Figure 1. Solar thermal reduction of $FeSO_4$ (Ar = 100 moles/sec).

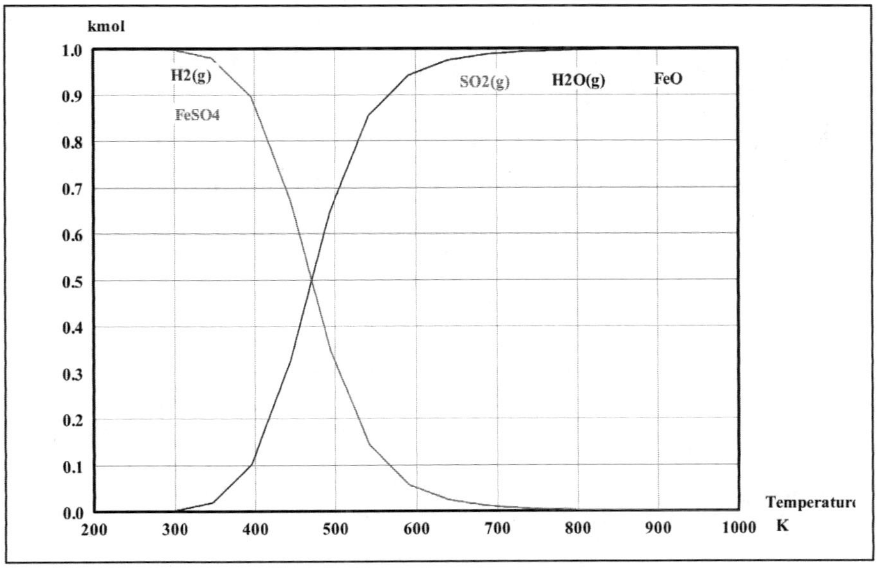

Figure 2. Water splitting reaction using IO-IS cycle (Ar = 100 moles/sec).

3. Exergy Analysis

The energy conversion efficiency of the solar thermochemical IO-IS water splitting cycle was determined by following the second law thermodynamic (exergy) analysis using the process flow diagram shown in Figure 4. This process is comprised of a solar reactor, FeO oxidizer (H_2 generator), theoretical H_2/O_2 ideal fuel cell, gas separator, and few heaters and coolers. Production of H_2 is carried out at 1 atm and steady state conditions with negligible viscous losses and changes in kinetic and potential energies. The energy penalties associated with the separation of SO_2 from O_2 and Ar are not accounted during the efficiency calculation. Also, the heat exchangers required for recovering the sensible latent heat are omitted from thermodynamic considerations. In addition, several other assumptions are made such as follows:

- The solar reactor is assumed to be a perfectly insulated blackbody absorber
- No convective/conductive heat losses
- Effective emissivity and absorptivity equal to unity
- All the products separate naturally without expending any work
- All reactions reach 100% completion

The exergy analysis was carried out by following the methodology and governing equations derived and used previously for other MO based solar thermochemical cycles [7]. Thermodynamic properties were extracted from HSC software and databases and the analysis is normalized to the $FeSO_4$ molar flow rate (1 mol/sec) entering the solar reactor.

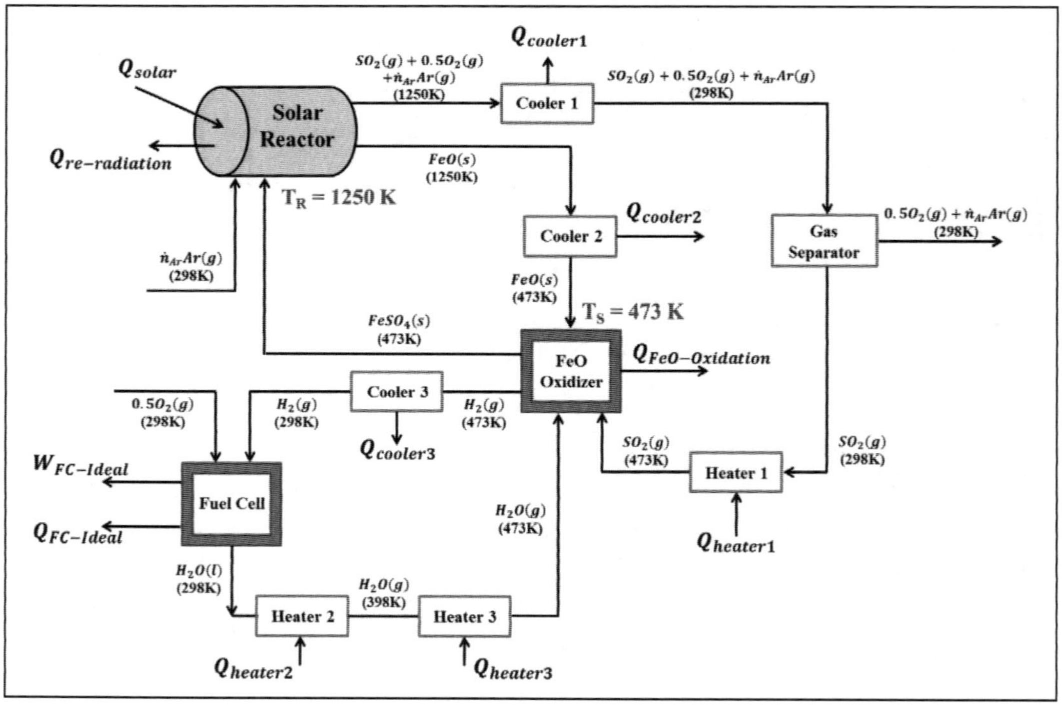

Figure 3. Process configuration for H_2 production via two-step solar thermochemical IO-IS water splitting cycle (Ar = 100 moles/sec).

3.1 *Solar Energy Absorption Efficiency* ($\eta_{absorption}$)

The solar energy absorption efficiency ($\eta_{absorption}$) is defined as the net rate at which energy is being absorbed by the solar reactor divided by the solar power input through the aperture of the solar reactor. The $\eta_{absorption}$ can be calculated as:

$$\eta_{absorption} = 1 - \left(\frac{\sigma T_R^4}{IC}\right) \qquad (3)$$

Where, σ = Stefan-Boltzmann constant, I = normal beam insolation $1(kW/m^2)$, C= solar flux concentration ratio of the solar concentrating system (5000 suns), $Q_{reactor-net}$= net energy absorbed in the solar reactor, Q_{solar} = solar energy input. For IO-IS water splitting cycle, at T_R= 1250K, the $\eta_{absorption}$ is observed to be equal to 97.23%.

3.2 *Net Solar Energy Absorbed for The Operation of IO-IS Cycle*

For the operation of solar thermochemical IO-IS water splitting cycle, following energy requirements associated with following steps needs to be fulfilled by solar energy input.

- Energy required for the complete reduction of $FeSO_4$ into FeO, SO_2, and O_2
- Energy required for heating of Ar from ambient conditions upto T_R
- Energy required for heating of SO_2 from ambient conditions upto T_S
- Energy required for the conversion of liquid water (298K) into steam (398K)
- Energy required for the heating of steam from 398K upto T_S

The net solar energy absorbed for the operation of solar thermochemical IO-IS water splitting cycle can be calculated as:

$$Q_{reactor-net} = Q_{FeSO_4-reduction} + Q_{Ar-heating} + Q_{heater-1} + Q_{heater-2} + Q_{heater-3} \quad (4)$$

3.2.1 $Q_{heater-1}$

The energy required for the heating of SO_2 from ambient conditions upto 298K is given by following equation (5) and $Q_{heater-1}$ is observed to be equal to 8.072 kW.

$$Q_{heater-1} = \dot{n}\Delta H|_{SO_2(g)@298K \rightarrow SO_2(g)@473K} \qquad (5)$$

3.2.2 $Q_{heater-2}$

The energy required for the production of steam via heating of the liquid water is 47.366 kW (according to Eq. 6).

$$Q_{heater-2} = \dot{n}\Delta H|_{H_2O(l)@298K \rightarrow H_2O(g)@398K} \qquad (6)$$

3.2.3 $Q_{heater-3}$

$$Q_{heater-3} = \dot{n}\Delta H|_{H_2O(g)@398K \rightarrow H_2O(g)@473K} \quad (7)$$

After steam generation, it is needed to heat the steam upto the splitting temperature (473K). The energy penalty associated with the heating of steam from 398K upto 473K can be estimated according to Eq. 7 and is observed to be equal to 2.625 kW.

3.2.4 $Q_{Ar-heating}$

To achieve 100% thermal reduction of $FeSO_4$ at lower temperatures of 1250K, 100 moles/sec of inert *Ar* flow rate was used as a carrier gas inside the solar reactor. Equation (8) gives the amount of energy needed to heat the inert *Ar* from ambient conditions upto thermal reduction temperature.

$$Q_{Ar-heating} = \dot{n}\Delta H|_{Ar(g)@298K \rightarrow Ar(g)@T_R} \quad (8)$$

After solving this equation it was realized that 1982.06 kW of heat energy is needed to heat the inert Ar from 298K upto 1250K.

3.2.5 $Q_{FeSO_4-reduction}$

The Energy required for the complete reduction of $FeSO_4$ (at 1250 K) into FeO, SO_2 (g), and O_2 (g) is given by following equation and observed to equal to 467.033 kW.

$$Q_{FeSO_4-reduction} = \dot{n}\Delta H|_{FeSO_4@473K \rightarrow FeO + SO_2(g) + {}^1/_2 O_2(g)@T_R} \quad (9)$$

At the end, the total net solar energy absorbed for the operation of IO-IS cycle is observed to be equal to 2507.156 kW (according to Eq. 4).

3.3 *Solar Energy Input To The Solar Reactor* (Q_{solar})

Based on the calculated $\eta_{absorption}$ and $Q_{reactor-net}$, the amount of solar energy input required to operate the solar reactor (Q_{solar}) dealing with the IO-IS water splitting thermochemical process can be determined by equation (10) and observed to be equal to 2578.58 kW.

$$Q_{solar} = \frac{Q_{reactor-net}}{\eta_{absorption}} \quad (10)$$

3.4 *Radiation Heat Losses from The Solar Reactor*

Radiation heat losses and % re-radiation from the solar reactor conducting the thermal reduction of $FeSO_4$ can be calculated by equations (11) and (12):

$$Q_{re-radiation} = Q_{solar} - Q_{reactor-net} \qquad (11)$$

$$\% \, re - radiation \, losses = \frac{Q_{re-radiation}}{Q_{solar}} \times 100 \qquad (12)$$

Computational thermodynamic modeling indicate that radiation heat losses = 71.246 kW and % re-radiation = 2.77%.

3.5 *Rate of Heat Rejected To The Surrounding By Cooler 1 And Cooler 2*

Gaseous (SO_2, O_2, and Ar) and solid (FeO) products generated during thermal reduction of FeSO4 exits the solar reactor and further cooled down via passing through cooler 1 and cooler 2. The rate of heat rejected by cooler 1 and cooler 2 is given by Eqs. (13) and (14).

$$Q_{cooler-1} = -\dot{n}\Delta H|_{SO_2(g)+0.5O_2(g)+Ar(g)@T_R \rightarrow SO_2(g)+0.5O_2(g)+Ar(g)@298K} \qquad (13)$$

$$Q_{cooler-2} = -\dot{n}\Delta H|_{FeO@T_R \rightarrow FeO@473K} \qquad (14)$$

It is observed that the heat energy released during the cooling of the gaseous products is higher as compared the heat energy released during the cooling of the solid FeO. For instance, $Q_{cooler-1}$ = 2053.39 Kw and $Q_{cooler-2}$= 47.786 kW, respectively.

3.6 *Rate of Heat Rejected To The Surrounding By FeO Oxidizer And Cooler 3*

FeO generated via thermal reduction of FeSO4 is transferred to the FeO oxidizer and reacted with the combined stream of SO_2 and H_2O at 473K for the production of H_2. By assuming 100% oxidation of FeO producing $FeSO_4$ and H_2 via water splitting reaction, the rate of heat rejected to the surrounding by FeO oxidizer is estimated as 118.43 kW according to equation (15).

$$Q_{FeO-oxidizer} = -\dot{n}\Delta H|_{FeO+H_2O(g)+SO_2(g)@473K \rightarrow FeSO_4+H_2(g)@473K} \qquad (15)$$

H_2 generated at 473K via water splitting reaction is allowed to pass through cooler 3 before entering the ideal fuel cell (operated at 298K). According to equation (16), 5.035 kW of heat is released by cooler 3 due to the cooling of H_2 from 473K to 298K.

$$Q_{cooler-3} = -\dot{n}\Delta H|_{H_2(g)@473K \rightarrow H_2(g)@298K} \qquad (16)$$

3.7 *Rate of Theoretical Work Performed And Heat Rejected By The Ideal Fuel Cell*

Solar thermochemical IO-IS water splitting cycle is completed with the addition of an ideal H_2/O_2 fuel cell with 100% work efficiency (assumption). This ideal fuel cell is added to the cycle to extract the maximum work from the net H_2 produced by this process. The rate of theoretical work performed by the ideal fuel cell is calculate by equation (17) and observed to be equal to

237.05 kW. Likewise, the rate of heat energy released by the ideal fuel cell is determined as equal to 48.56 kW [according to equation (18)].

$$W_{FC-Ideal} = -\dot{n}\,\Delta G|_{H_2(g)+0.5O_2(g)@298K \rightarrow H_2O(l)@298K} \qquad (17)$$

$$Q_{FC-Ideal} = -(298) \times \dot{n}\,\Delta S|_{H_2(g)+0.5O_2(g)@298K \rightarrow H_2O(l)@298K} \quad (18)$$

3.8 Solar To Fuel Conversion Efficiency

Cycle efficiency of the solar thermochemical IO-IS water splitting process is defined as the ratio of theoretical work performed by the ideal fuel cell to the solar energy input:

$$\eta_{cycle} = \frac{W_{FC-Ideal}}{Q_{solar}} \qquad (14)$$

Furthermore, the solar-to-fuel energy conversion efficiency of the solar thermochemical SO-SS water splitting process is defined as the ratio of higher heating value (HHV) of the H_2 produced to the solar energy input:

$$\eta_{solar-to-fuel} = \frac{HHV_{H_2}}{Q_{solar}} \qquad (15)$$

Where,

$$HHV_{H_2} = -\dot{n}\,\Delta H|_{H_2(g)+0.5O_2(g)@298K \rightarrow H_2O(l)@298K} \qquad (16)$$

As per the HSC computation thermodynamic modeling, η_{cycle} and $\eta_{solar-to-fuel}$ equal to 9.193% and 11.05%, respectively.

4. Conclusion

Two-step solar thermochemical cycles for H_2 production via IO-IS based H_2O splitting reactions have been thermodynamically examined. Thermodynamic equilibrium calculations performed at 1 atm pressure and in absence of inert Ar indicate that the H_2 production via water splitting reaction is feasible below 473 K. Similarly, the thermodynamic modeling also indicates that the thermal reduction of $FeSO_4$ into FeO, SO_2 and O_2 can be achieved at 1250 K when 100 mol/sec of inert Ar flow rate is used. The second law thermodynamic analysis indicate that indicate $_{absorption}$ = 97.23%, Q_{solar}= 2578.58 kW, $Q_{reactor-net}$ = 2507.16 kW, $Q_{re-radiation}$ = 71.25 kW, $Q_{FeO-oxd}$ = 118.43 kW, and η_{cycle}= 9.193%, and $\eta_{solar-to-fuel}$= 11.05%, respectively.

5. Acknowledgments

The authors gratefully acknowledge the financial support provided by the Qatar University Internal Grants QUUG-CENG-CHE-13/14-4 and QUUG-CENG-CHE-14\15-10.

References

[1] Agrafiotis C, Pagkoura C, Lorentzou S, Kostoglou M, Konstandopoulos AG. Hydrogen production in solar reactors. Catalysis Today 2007;127:265–77.

[2] Steinfeld A, Sanders S, Palumbo R. Design aspects of solar thermochemical engineering – A case study: two-step water-splitting cycle using the Fe3O4/FeO redox system. Solar Energy 1999;43:43–53.

[3] Kodama T, Nakamuro Y, Mizuno T. A two-step thermochemical water spitting by iron-oxide on stabilized zirconia. Journal of Solar Energy Engineering 2006;128:3–7.

[4] Gokon N, Murayama H, Umeda J, Hatamachi T, Kodama T. Monoclinic zirconia-supported Fe3O4 for the two-step water-splitting thermochemical cycle at high thermal reduction temperatures of 1400–1600°C. International Journal of Hydrogen Energy 2009;34:1208–17.

[5] Ishihara H, Kaneko H, Hasegawa N, Tamaura Y. Two-step water-splitting at 1273–1623K using yttria-stabilized zirconia-iron oxide solid solution via co-precipitation and solid-state reaction. Energy 2008;33:1788–93.

[6] Miller J, Allendorf M, Diver R, Evans L, Seigel N, Stuecker J. Metal oxide composites and structures for ultra-high temperature solar thermochemical cycles. Journal of Material Science 2008;43:4714–28.

[7] Steinfeld A. Solar hydrogen production via a two-step water-splitting thermochemical cycle based on Zn/ZnO redox reactions. International Journal of Hydrogen Energy 2002;37:611–19.

[8] Abanades S. Thermogravimetric analysis of CO2 and H2O reduction from solar nanosized Zn powder for thermochemical fuel production. Industrial & Engineering Chemistry Research 2012;51:741–50.

[9] Ernst F, Steinfeld A, Pratsinis S. Hydrolysis rate of submicron Zn particles for solar H2 synthesis. International Journal of Hydrogen Energy 2009;34:1166–75.

[10] Vishnevetsky I, Epstein M. Production of hydrogen from solar zinc in steam atmosphere. International Journal of Hydrogen Energy 2007;32:2791–2802.

[11] Abanades S, Charvin P, Lemont F, Flamant G. Novel two-step SnO2/SnO water-splitting cycle for solar thermochemical production of hydrogen. International Journal of Hydrogen Energy 2008;33:6021–30.

[12] Charvin P, Abanades S, Lemont F, Flamant G. Experimental study of SnO2/SnO/Sn thermochemical systems for solar production of hydrogen. AIChE Journal 2008;54:2759–67.

[13] Vishnevetsky I, Epstein M. Tin as a possible candidate for solar thermochemical redox process for hydrogen production. Journal of Solar Energy Engineering 2009;131:021007-1–021007-8.

[14] Scheffe JR, Li J, Weimer AW, A spinel ferrite/hercynite water-splitting redox cycle. International Journal of Hydrogen Energy 2010;35:3333–3340.

[15] Bhosale RR, Shende RV, Puszynski JA, H2 generation from thermochemical water-splitting using sol-gel derived Ni-ferrite. Journal of Energy and Power Engineering 2010;4:27–38.
[16] Neises M, Roeb M, Schmucker M, Sattler C, Pitz-Paal R. Kinetic investigations of the hydrogen production step of a thermochemical cycle using mixed iron oxides coated on ceramic substrates. International Journal of Energy Research 2010;34:651–61.
[17] Bhosale RR, Shende RV, Puszynski JA, H2 generation from thermochemical water-splitting using sol-gel synthesized Zn/Sn/Mn-doped Ni-ferrite. International Review of Chemical Engineering 2010;2:852–62.
[18] Lorentzou S, Zygogianni A, Tousimi K, Agrafiotis C, Konstandopoulos AG. Advanced synthesis of nanostructured material for environmental applications. Journal of Alloys and Compounds 2009;483:302–05.
[19] Bhosale RR, Khadka RP, Shende RV, Puszynski JA. H2 generation from two-step thermochemical water-splitting reaction using sol-gel derived SnxFeyOz. Journal of Renewable and Sustainable Energy 2011;3:063104-1–063104-16.
[20] Varsano F, Padella F, Alvani C, Bellusci M, Barbera A. Chemical aspects of the water-splitting thermochemical cycle based on sodium manganese ferrite. International Journal of Hydrogen Energy 2012;37:11595–601.
[21] Bhosale RR, Shende RV, Puszynski JA. Thermochemical water-splitting for H2 generation using sol-gel derived Mn-ferite in a packed bed reactor. International Journal of Hydrogen Energy 2012;37:2924–34.
[22] Scheffe JR, Steinfeld A. Thermodynamic analysis of cerium-based oxides for solar thermochemical fuel production. Energy & Fuels 2012;26:1928–36.
[23] Abanades S, Flamant G. Thermochemical hydrogen production from a two-step solar-driven water-splitting cycle based on cerium oxides. Solar Energy 2006;80:1611–23.
[24] Gal A, Abanades S. Dopant incorporation in ceria for enhanced water-splitting activity during solar thermochemical hydrogen generation. Journal of Physical Chemistry C 2012;116:13516–523.
[25] Lin X, Zhang Y, Wang Z, Wang R, Zhou J, Cen K. Hydrogen production by HI decomposition over nickel–ceria–zirconia catalysts via the sulfur–iodine thermochemical water-splitting cycle. Energy Conversion and Management 2014;84:664–70.
[26] Murphy IV J, O'Connell J. Process simulations of HI decomposition via reactive distillation in the sulphur-iodine cycle for hydrogen manufacture. International Journal of Hydrogen Energy 2012;37:4002–11.
[27] Dehghani S, Sayyaadi H. Energy and exergetic evaluations of Bunsen section of the sulfur–iodine thermochemical hydrogen production plant. International Journal of Hydrogen Energy 2013;38:9074–84.
[28] Lin S, Flaherty R. Design studies of the sulphur trioxide decomposition reactor for the sulphur cycle hydrogen production process. International Journal of Hydrogen Energy 1983;8:589–96.
[29] Hinkley J, O'Brein J, Fell C, Lindquist S. Prospects for solar only operation of the hybrid sulphur cycle for hydrogen production. International Journal of Hydrogen Energy 2011;8:589–96.

15AIChE
Spring Meeting, Austin, TX

Computational Modeling of Mixed Solids for CO_2 Capture Sorbents

Yuhua Duan
National Energy Technology Laboratory, United States Department of Energy
626 Cochrans Mill Road, Pittsburgh, Pennsylvania 15236, USA
Yuhua.duan@netl.doe.gov

Prepared for Presentation at
American Institute of Chemical Engineers
2015 Spring Meeting
11th Global Congress on Process Safety
Austin, Texas
April 27-30, 2015

UNPUBLISHED

Computational Modeling of Mixed Solids for CO_2 Capture Sorbents

Yuhua Duan
National Energy Technology Laboratory, United States Department of Energy
626 Cochrans Mill Road, Pittsburgh, Pennsylvania 15236, USA
Yuhua.duan@netl.doe.gov

Keywords: CO_2 solid sorbents, *ab initio* thermodynamics, first-principles density functional theory, lattice phonon dynamics, computational screening

Abstract

Since current technologies for capturing CO_2 to fight global climate change are still too energy intensive, there is a critical need for development of new materials that can capture CO_2 reversibly with acceptable energy costs. Accordingly, solid sorbents have been proposed to be used for CO_2 capture applications through a reversible chemical transformation. By combining thermodynamic database mining with first principles density functional theory and phonon lattice dynamics calculations, a theoretical screening methodology to identify the most promising CO_2 sorbent candidates from the vast array of possible solid materials has been proposed and validated. The calculated thermodynamic properties of different classes of solid materials versus temperature and pressure changes were further used to evaluate the equilibrium properties for the CO_2 adsorption/desorption cycles. According to the requirements imposed by the pre- and post- combustion technologies and based on our calculated thermodynamic properties for the CO_2 capture reactions by the solids of interest, we were able to screen only those solid materials for which lower capture energy costs are expected at the desired pressure and temperature conditions. Only those selected CO_2 sorbent candidates were further considered for experimental validations. The *ab initio* thermodynamic technique has the advantage of identifying thermodynamic properties of CO_2 capture reactions without any experimental input beyond crystallographic structural information of the solid phases involved. Such methodology not only can be used to search for good candidates from existing database of solid materials, but also can provide some guidelines for synthesis new materials. In this presentation, we apply our screening methodology to mixing solid systems to adjust the turnover temperature to help on developing CO_2 capture Technologies.

1. Introduction

Nowadays, the burning of fossil fuels is still the main energy source for the world's economy. One consequence of the use of these fuels is the emission of huge quantities of CO_2 into the atmosphere creating environmental problems such as global climate warming.[1-5] In order to solve such problems, there is a need to reduce CO_2 emissions into atmosphere by capturing and sequestrating CO_2.[4, 6, 7] One approach to solve such environmental problems is to capture and sequester the CO_2. Current technologies for capturing CO_2 including solvent-based (amines) and CaO-based materials are still too energy intensive. Hence, there is critical need for development of new materials that can capture and release CO_2 reversibly with acceptable energy costs. In particular, solid oxide sorbent materials have been proposed for capturing CO_2

through a reversible chemical transformation leading primarily to formation of carbonate products. Solid sorbents containing alkali and alkaline earth metals have been reported in several previous studies to be promising candidates for CO_2 sorbent applications due to their high CO_2 absorption capacity at moderate working temperatures.[8-11]

During past few years we developed a theoretical methodology to identify promising solid sorbent candidates for CO_2 capture by combining thermodynamic database searching with *ab initio* thermodynamics obtained based on first-principles density functional theory (DFT) and lattice phonon dynamics.[8-10, 12-16] As shown in Fig.1, the primary outcome of our screening scheme is a list of promising CO_2 sorbents with optimal energy usage.

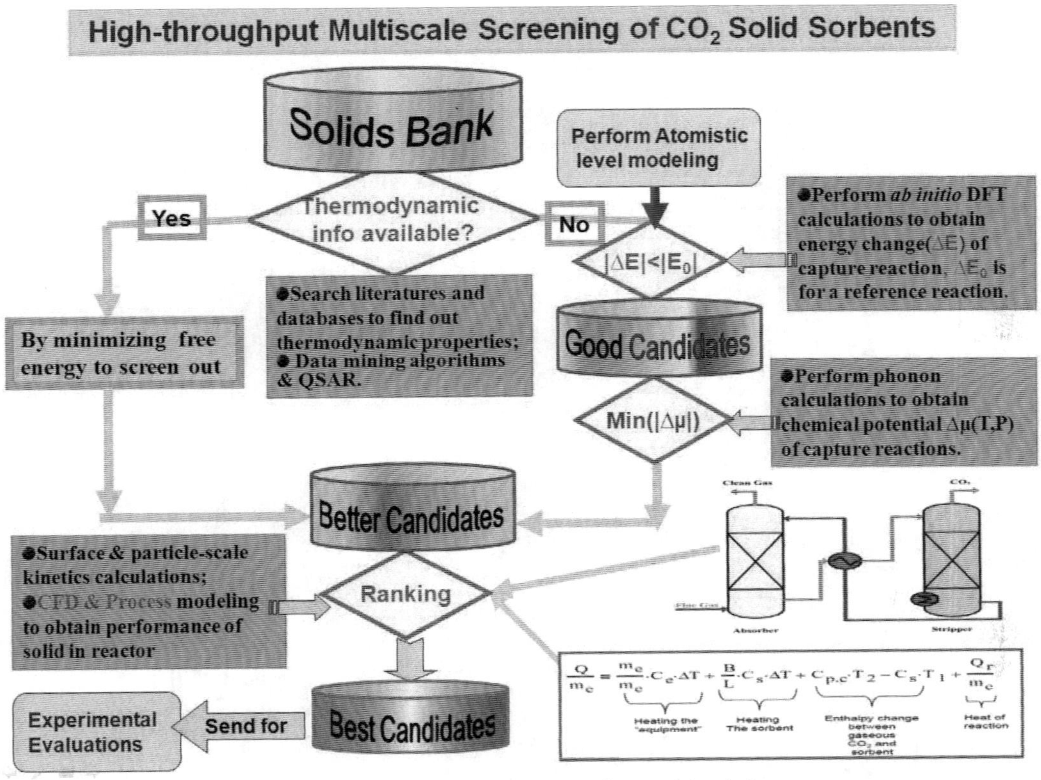

Fig. 1. Schematic of our screening methodology.

At a given CO_2 pressure, the turnover temperature (T_t) of an individual solid capture CO_2 reaction is fixed. Such T_t may be outside the operating temperature range (ΔT_o) for a particularly capture technology. In order to adjust T_t to fit the practical ΔT_o, in this study, we demonstrate that by mixing different types of solids it's possible to shift T_t to the range of practical operating conditions.

2. Calculation Methods for Mixed Solid Sorbents

The complete description of the computational methodology together with relevant applications can be found in our previous publications.[8-10, 12-19] The CO_2 capture reactions of solids can be expressed generically in the form (for convenient description, we normalized the reaction to 1 mole of CO_2)

$$\sum_{Ri} n_{Ri} Solid\_R_i + CO_2 \leftrightarrow \sum_{Pj} n_{Pj} Solid\_P_j \qquad (a)$$

where n_{Ri}, n_{Pj} are the numbers of moles of reactants (R_i) and products (P_j) involved in the capture reactions. We treat the gas phase CO_2 as an ideal gas. By assuming that the difference between the chemical potentials ($\Delta\mu^o$) of the solid phases of reactants (R_i) and products (P_j) can be approximated by the difference in their total energies (ΔE^{DFT}), obtained directly from DFT calculations, and the vibrational free energy of the phonons and by ignoring the PV contribution terms for solids, the variation of the chemical potential ($\Delta\mu$) for reaction (a) with temperature and pressure can be written as[8-10, 12-19]

$$\Delta\mu(T, P) = \Delta\mu^0(T) - RT \ln \frac{P_{CO_2}}{P_0} \qquad (1)$$

where

$$\Delta\mu^0(T) \approx \Delta E^{DFT} + \Delta E_{ZP} + \Delta F^{PH}(T) - G^0_{CO_2}(T) \qquad (2)$$

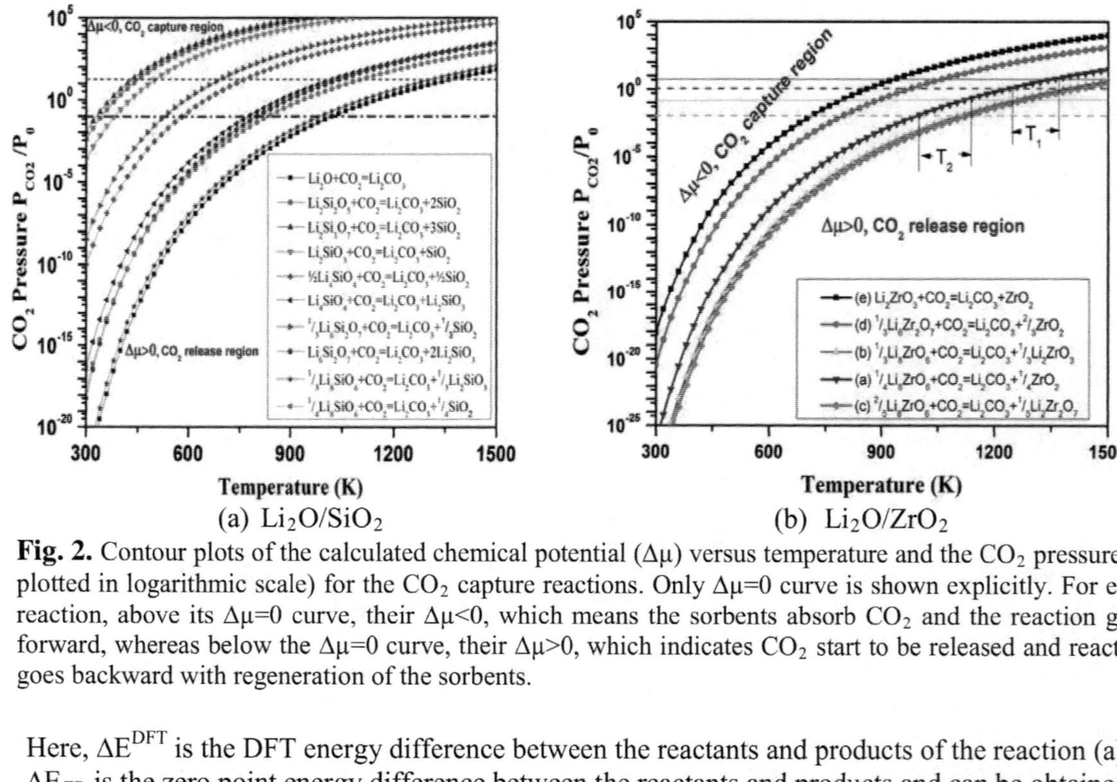

(a) Li_2O/SiO_2 (b) Li_2O/ZrO_2

Fig. 2. Contour plots of the calculated chemical potential ($\Delta\mu$) versus temperature and the CO_2 pressure (*P* plotted in logarithmic scale) for the CO_2 capture reactions. Only $\Delta\mu$=0 curve is shown explicitly. For each reaction, above its $\Delta\mu$=0 curve, their $\Delta\mu<0$, which means the sorbents absorb CO_2 and the reaction goes forward, whereas below the $\Delta\mu$=0 curve, their $\Delta\mu>0$, which indicates CO_2 start to be released and reaction goes backward with regeneration of the sorbents.

Here, ΔE^{DFT} is the DFT energy difference between the reactants and products of the reaction (a), ΔE_{ZP} is the zero point energy difference between the reactants and products and can be obtained directly from phonon calculations. ΔF^{PH} is the phonon free energy change excluding zero-point energy (which is already counted into the ΔE_{ZP} term) between the solids of products and reactants. P_{CO2} is the partial pressure of CO_2 in the gas phase and P_0 is the standard state reference pressure taken to be 1 bar. The heat of reaction ($\Delta H^{cal}(T)$) can be evaluated through the following equation

$$\Delta H^{cal}(T) = \Delta\mu^0(T) + T[\Delta S_{PH}(T) - S_{CO_2}(T)] \qquad (3)$$

where $\Delta S_{PH}(T)$ is the difference of entropies between product solids and reactant solids. The free energy of CO_2 (G^0_{CO2}) can be obtained from standard statistical mechanics,[9, 10, 16] and its entropy (S_{CO2}) can be found in the empirical thermodynamic databases.[20]

3. Results and Discussions

For a given CO_2 capture process, its optimal working conditions (CO_2 pressures of pre- and after-capture, absorption/desorption temperature range (ΔT_o), *etc.*) were fixed. However, at a given CO_2 pressure, the turnover temperature (T_t) of an individual solid capture CO_2 reaction is fixed. Such T_t may be outside the operating temperature range ΔT_o for a particularly capture technology. In order to adjust T_t to fit the practical working through reversible chemical transformations ΔT_o, we have demonstrated that by mixing different types of solids it's possible to shift T_t to the practical operating ΔT_o range.[8, 21, 22] Generally, when we mix two solids *A* and *B* to form a new sorbent *C*, the turnover temperature of the newly resulted system (T_C) is located between those of *A* and *B* (T_A, T_B). Here it was assumed that *A* is a strong CO_2 sorbent while *B* is a weak CO_2 sorbent and $T_A > T_B$. Also, we assumed that the desired operating temperature T_O is between T_A and T_B ($T_A > T_O > T_B$). Depending on the properties of *A* and *B*, we have typically three scenarios to synthesize the mixing sorbent *C*:

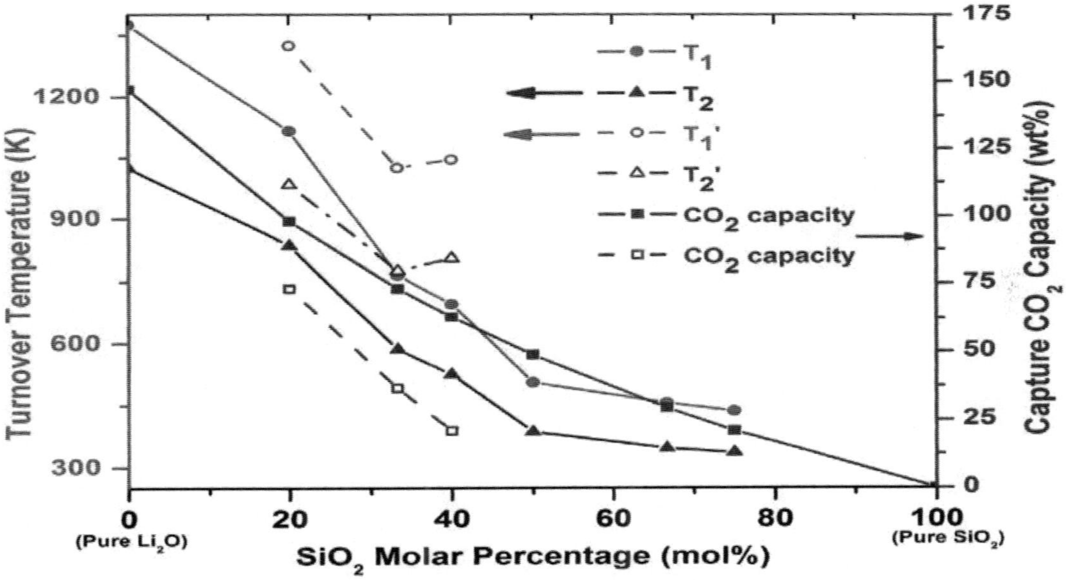

Fig. 3 The dependence of the turnover temperatures defined in the text and of CO_2 capture capacity on molar percentage of SiO_2 in the silicates for which calculations are reported here. T_1 and T_1' are the turnover temperatures under pre-combustion conditions with CO_2 partial pressure at 20 bars, while T_2 and T_2' are the turnover temperatures under post-combustion conditions with CO_2 partial pressure at 0.1 bar. The solid lines indicate to convert lithium silicates to SiO_2 and Li_2CO_3 (T_1, T_2). For those Li_2O-rich lithium silicates (Li_8SiO_6, Li_4SiO_4, $Li_6Si_2O_7$) capturing CO_2, the data shown in dash lines indicate convert them to Li_2SiO_3 and Li_2CO_3 (T_1', T_2'). The corresponding CO_2 capture capacities are plotted with open blue squares.

3.1. $T_A >> T_B$ and the *A* component is the key part to capture CO_2.

An example of this case is represented by Li_2O. This is a very strong CO_2 sorbent which forms Li_2CO_3. However, its regeneration from Li_2CO_3 only can occur at very high temperature (T_A). In order to move its T_A to lower temperatures, one can mix some weak CO_2 sorbents (such as SiO_2, ZrO_2). Fig. 2 shows the relationship of chemical potential, P_{CO2}, and T of the CO_2 capture reactions by the mixed Li_2O/SiO_2 and Li_2O/ZrO_2 solids with different mixing ratios. Fig. 3 shows the turnover temperature and the CO_2 capture capacity of Li_2O/SiO_2 mixture versus the ratio of Li_2O/SiO_2.[8-10, 12, 18, 21, 23, 24]

From Figs. 2 and 3, one can see that after mixing Li_2O and SiO_2 (or ZrO_2) with different Li_2O/SiO_2 (or Li_2O/ZrO_2) ratios, the T_C of the newly formed *C* compound (silicate or zirconate) is lower than T_A of pure Li_2O and could be close to the ΔT_o range to fit the practical needs.

3.2 $T_A >> T_B$ and B component is the key part to capture CO_2

In this case, since T_B is lower than T_O, by mixing *A* into *B* will increase the turnover temperature T_C of the *C* solid to values closer to T_o. For example, pure MgO has a very high theoretical CO_2 capture capacity. However, its turnover temperature (250 °C) is lower than the required temperature range of 300-470 °C used in warm gas clean up technology and its practical CO_2 capacity is very low, and therefore, it cannot be used directly as a CO_2 sorbent in this technology.[25-27] As shown in Fig. 4, by mixing alkali metal oxides M_2O (M=Na, K, Cs, Ca) or carbonates (M_2CO_3) into MgO, the corresponding newly formed mixing systems have higher turnover temperatures making them useful as CO_2 sorbents through the reaction $MgO + CO_2 + M_2CO_3 = M_2Mg(CO_3)_2$.[25, 28]

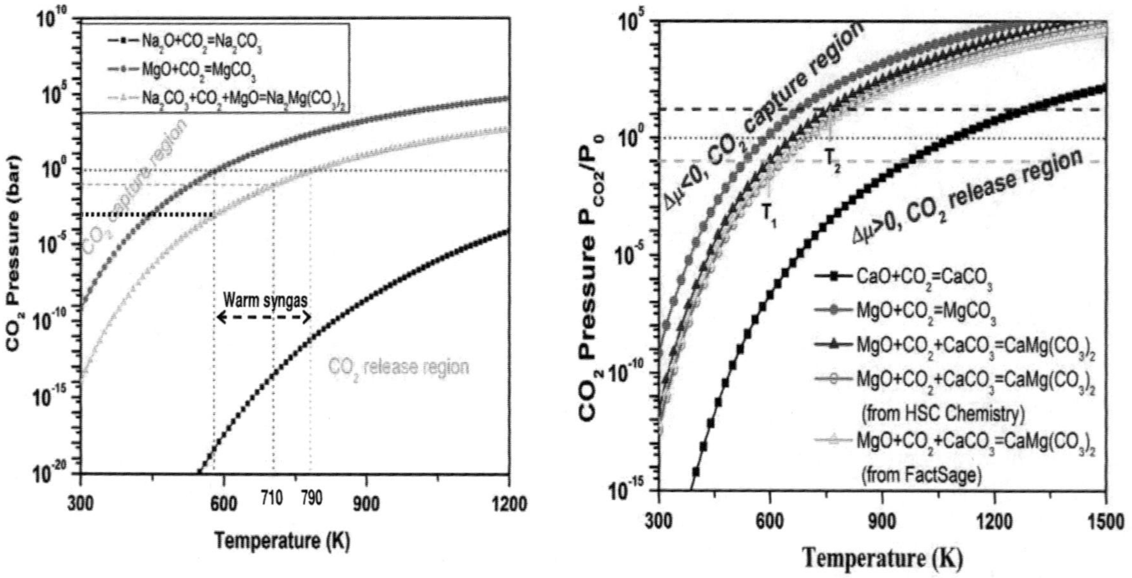

(a) MgO/Na_2CO_3 (Na_2O) (b) $MgO/CaCO_3$ (CaO)

Fig. 4. Plots of the calculated chemical potentials versus CO_2 pressures and temperatures for the CO_2 capture reaction by MgO to form double salts.

3.3 T_A and T_B are close to each other and both A and B are active for CO_2 capture

In this case, both *A* and *B* components are active to capture CO_2, and the CO_2 capacity of the mixture is the summation of those of *A* and *B*. As we know another potential advantage of mixing solids is to increase the surface area of the solids in order to have faster reaction rates. Such a mixing scenario doesn't show too much advantage in shifting the capture temperature, but may enhance the kinetics of the capture process and eventually make the mixtures more efficient. Although there is no such report in literature, we think such an attempt is worthwhile and are working on several doped systems. Figure 5 shows a case of Na-/K- doped Li_2ZrO_3 capture CO_2 through the following reaction (b):[29]

$$LiMZrO_3 + CO_2 \leftrightarrow \frac{1}{2}Li_2CO_3 + \frac{1}{2}M_2CO_3 + ZrO_2 \qquad (M = Na, K) \qquad \text{(b)}$$

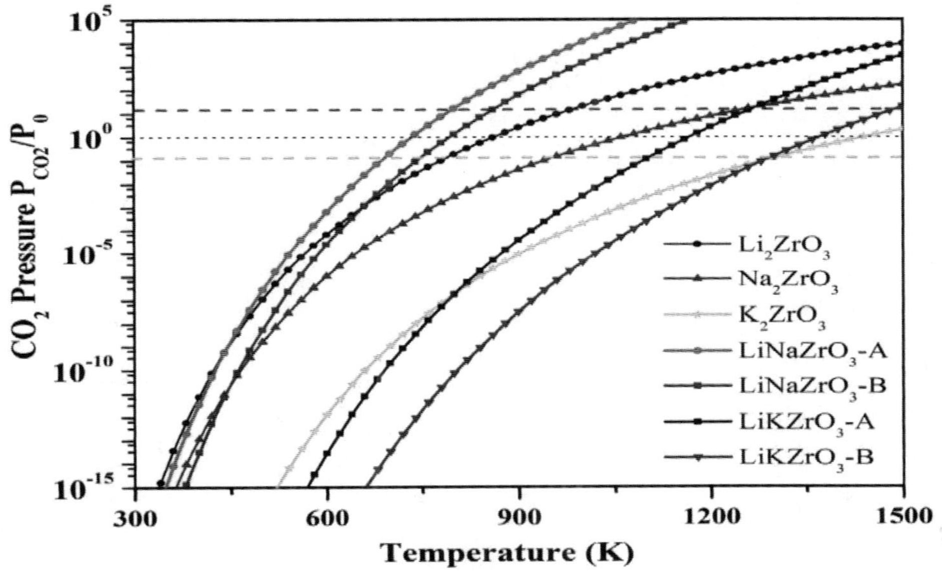

Fig. 5. The contour plotting of calculated chemical potentials versus CO_2 pressures and temperatures of the CO_2 capture reactions by $LiMZrO_3$ and M_2ZrO_3. Y-axis plotted in logarithm scale. Only the $\Delta\mu=0$ curve is shown explicitly. For each reaction, above its $\Delta\mu=0$ curve, their $\Delta\mu<0$, which means the solids absorb CO_2 and the reaction goes forward, whereas below the $\Delta\mu=0$ curve, their $\Delta\mu>0$, which means the CO_2 start to release and the reaction goes backward to regenerate the sorbents.

All of the reactions are thermodynamically favorable for CO_2 sorption over a quite wide range of temperatures and reduced CO_2 pressures. But a well preforming CO_2 solid sorbent should not only easily absorb CO_2 in the first half cycle but also easily release CO_2 from the product (M_2CO_3 and ZrO_2 for example) in the second half cycle. Table 1 shows the calculated thermodynamical properties and turnover temperatures of $LiMZrO_3$ capture CO_2 reaction (b).

From Fig.5 and Table 1, one can see that they are not good sorbents for capturing CO_2 for pre-combustion technology. However, they could be used for high-temperature post-combustion CO_2 capture with T_2=685K, 745 K for $LiNaZrO_3$ (phases _A and _B) and T_2=1095K, 1285 K for $LiKZrO_3$ (phases _A and _B) respectively. When compared to Li_2ZrO_3, the K substituted system

$LiKZrO_3$ does not decrease the turnover temperature of the CO_2 capture reaction. In other words, although the $LiKZrO_3$ has a stronger driving force to capture CO_2, with improved CO_2 capture capability when compared to Li_2ZrO_3, its regeneration only can happen at higher temperature. Therefore, $LiKZrO_3$ is not a good CO_2 sorbent, since it requires more energy to regenerate. In contrast, the Na-substituted sorbent system $LiNaZrO_3$ does have a more compatible turn over temperature, making it a possible good candidate as a CO_2 sorbent. Such conclusions are in good agreement with experimental findings.[30-32]

In summary, from the calculated thermodynamic properties of $LiMZrO_3$ reacting with CO_2, we found that the CO_2 sorbent performance of $LiNaZrO_3$ is better than that of Li_2ZrO_3, but the performance of $LiKZrO_3$ is much worse since its regeneration temperature is higher. Therefore, the Na doped Li_2ZrO_3, $LiNaZrO_3$, is a better CO_2 sorbent for post-combustion capture technology. Our calculated thermodynamic results showed that the K doped Li_2ZrO_3, $LiKZrO_3$, didn't gain any improvements over Li_2ZrO_3 on its overall CO_2 capture performance.

Table 1. The weight percentage of CO_2 capture, the calculated energy change ΔE^{DFT}, the zero-point energy changes ΔE^{ZP} and the thermodynamic properties (ΔH, ΔG) of the CO_2 capture reactions. (unit: kJ/mol). The turnover temperatures (T_1 and T_2) of the reactions of CO_2 capture by solids under the conditions of pre-combustion (P_{CO2}=20 bar) and post-combustion (P_{CO2}=0.1 bar) are also listed.

reaction	absorbing CO_2 Wt%	ΔE^{DFT}	ΔE^{ZP}	ΔH (T=300K)	ΔG (T=300K)	Turnover T (K)	
						T_1	T_2
$Li_2ZrO_3+CO_2 \leftrightarrow Li_2CO_3+ZrO_2$ [a]	28.75	-146.648	11.311	-158.562	-103.845	1000	780
$LiNaZrO_3$_A$+CO_2 \leftrightarrow ½(Li_2CO_3+Na_2CO_3)+ZrO_2$	26.01	-152.936	7.069	-176.666	-110.892	805	685
$LiNaZrO_3$_B$+CO_2 \leftrightarrow ½(Li_2CO_3+Na_2CO_3)+ZrO_2$	26.01	-167.872	6.934	-191.526	-126.477	865	745
$LiKZrO_3$_A$+CO_2 \leftrightarrow ½(Li_2CO_3+K_2CO_3)+ZrO_2$	23.75	-264.115	6.006	-287.513	-225.611	1275	1095
$LiKZrO_3$_B$+CO_2 \leftrightarrow ½(Li_2CO_3+K_2CO_3)+ZrO_2$	23.75	-311.604	7.080	-332.612	-272.410	hT[b]	1285
$Na_2ZrO_3+CO_2 \leftrightarrow Na_2CO_3+ZrO_2$ [a]	23.76	-140.862	2.236	-158.327	-114.121	1275	925
$K_2ZrO_3+CO_2 \leftrightarrow K_2CO_3+ZrO_2$ [a]	20.24	-223.158	5.813	-238.490	-187.884	hT[b]	1285

[a] from Refs.[18, 23, 24].
[b] hT means the temperature is higher than our temperature range (1500K)

With similar approach, we have investigated the electronic structural and phonon properties of $Na_{2-\alpha}M_{\alpha}ZrO_3$ (M=Li, K, α=0, 0.5, 1.0, 1.5, 2.0) by density functional theory and first-principles determination of phonon dynamics.[33] The thermodynamic properties of CO_2 absorption/desorption in these materials were analyzed. The obtained results are presented in Fig.6 and Table 2.

Based on the calculated relationships among the chemical potential change, CO_2 pressure and temperature for CO_2 capture reactions by $Na_{2-\alpha}M_{\alpha}ZrO_3$ and thermogravimetric analysis experimental measurements, compared to pure Na_2ZrO_3, the Li- and K-doped mixtures $Na_{2-\alpha}M_{\alpha}ZrO_3$ have lower turnover temperatures (T_t) and higher CO_2 capacities. The calculated results show that the shift in T_t depends not only on the doping element, but also depends on the doping level. The Li-doped systems have larger T_t decreases than the K-doped systems. When increasing the Li-doping level α, the T_t of corresponding mixture $Na_{2-\alpha}Li_{\alpha}ZrO_3$ decreases further to a low temperature range. However in the case of K-doped systems $Na_{2-\alpha}K_{\alpha}ZrO_3$, although initial doping of K into Na_2ZrO_3 can shift its T_t to lower temperature range, further increasing the K-doping level α results in an increase in T_t. Therefore, compared to K-doing, lithium inclusion into Na_2ZrO_3 structure has a larger influence on the CO_2 capture performance.

Table 2. The weight percentage of CO_2 capture, the ratios of Na_2O:M_2O:ZrO_2 (M=Li, K), the calculated energy change ΔE^{DFT}, the zero-point energy changes ΔE^{ZP} and the thermodynamic properties (ΔH, ΔG) of the CO_2 capture reactions. (unit: kJ/mol). The turnover temperatures (T_1 and T_2) of the reactions of CO_2 capture by solids under the conditions of pre-combustion (P_{CO2}=20 bar) and post-combustion (P_{CO2}=0.1 bar) are also listed.[33]

reaction	absorbing CO_2 wt%	Na_2O:M_2O:ZrO_2 ratio	ΔE^{DFT}	ΔE^{ZP}	ΔH (T=300K)	ΔG (T=300K)	Turnover T (K)	
							T_1	T_2
$Na_2ZrO_3+CO_2 \leftrightarrow Na_2CO_3+ZrO_2$ [a]	23.76	1:0:1	-140.862	2.236	-158.327	-114.121	1275	925
$Na_{1.5}Li_{0.5}ZrO_3$-B+$CO_2 \leftrightarrow {}^3/_4Na_2CO_3+{}^1/_4Li_2CO_3+ZrO_2$	24.83	$^3/_4$:$^1/_4$:1	-170.881	4.667	-242.090	-159.144	805	715
$Na_{1.0}Li_{1.0}ZrO_3$-B+$CO_2 \leftrightarrow {}^1/_2Na_2CO_3+{}^1/_2Li_2CO_3+ZrO_2$	26.01	½:$^1/_2$:1	-157.839	6.480	-228.381	-142.555	745	675
$Na_{0.5}Li_{1.5}ZrO_3$-A+$CO_2 \leftrightarrow {}^1/_4Na_2CO_3+{}^3/_4Li_2CO_3+ZrO_2$	27.31	¼:$^3/_4$:1	-169.827	9.652	-237.765	-146.230	735	665
$Na_{1.5}K_{0.5}ZrO_3$-A+$CO_2 \leftrightarrow {}^3/_4Na_2CO_3+{}^1/_4K_2CO_3+ZrO_2$	22.77	$^3/_4$:$^1/_4$:1	-210.081	2.486	-281.253	-199.996	915	825
$Na_{1.0}K_{1.0}ZrO_3$-B+$CO_2 \leftrightarrow {}^1/_2Na_2CO_3+{}^1/_2K_2CO_3+ZrO_2$	21.86	½:$^1/_2$:1	-245.436	2.058	-316.736	-236.789	1015	915
$Na_{0.5}K_{1.5}ZrO_3$-B+$CO_2 \leftrightarrow {}^1/_4Na_2CO_3+{}^3/_4K_2CO_3+ZrO_2$	21.02	¼:$^3/_4$:1	-278.147	1.519	-349.077	-272.038	1125	1015
$Li_2ZrO_3+CO_2 \leftrightarrow Li_2CO_3+ZrO_2$ [b]	28.75	0:1:1	-146.648	11.311	-158.562	-103.845	1000	780
$K_2ZrO_3+CO_2 \leftrightarrow K_2CO_3+ZrO_2$ [a]	20.24	0:1:1	-223.158	5.813	-238.490	-187.884	hT[c]	1285

[a] From Ref.[24].
[b] From Ref.[18]
[c] hT means the temperature is higher than 1500K

All these results may be of great interest in the development of specific CO_2 capture applications. As it has been shown, the $Na_{2-\alpha}Li_{\alpha}ZrO_3$ and $Na_{2-\alpha}K_{\alpha}ZrO_3$ compositions can produce modifications in the CO_2 capture temperatures, which may be used in the design of a specific composition depending on the temperature range that industry requires. Our work has identified that the capture of CO_2 in zirconate materials is not simply a matter of substitutional element, but also the doping level. This insight will need to be considered during future sorbent development. We have also demonstrated that computational methods can be used to accurately predict aspects of CO_2 capture and have the potential to drive future work by identifying the most promising candidate materials.

4. Conclusions

The obtained results showed that by changing the mixing ratio of solid *A* and solid *B* to form mixed solid *C* it's possible to shift the turnover T_t of the newly formed solid *C* to fit the practical CO_2 capture technologies. When mixing SiO_2 or ZrO_2 into the strong Li_2O sorbent, one can obtain a series of lithium silicates (or zirconates) with T_t lower than that of pure Li_2O. By mixing oxides (Na_2O, K_2O, CaO) or their corresponding carbonates into MgO, the obtained mixtures exhibit different thermodynamic behaviors and their T_t are higher than that of pure MgO. Such results can be used to provide insights for designing new CO_2 sorbents. Therefore, although one single material taken in isolation might not be an optimal CO_2 sorbent to fit the particular needs to operate at specific temperature and pressure conditions, by mixing or doping two or more materials to form a new material, our results showed that it is possible to synthesize new CO_2 sorbent formulations which can fit the industrial needs. Our results also show that computational modeling can play a decisive role for identifying materials with optimal performance.

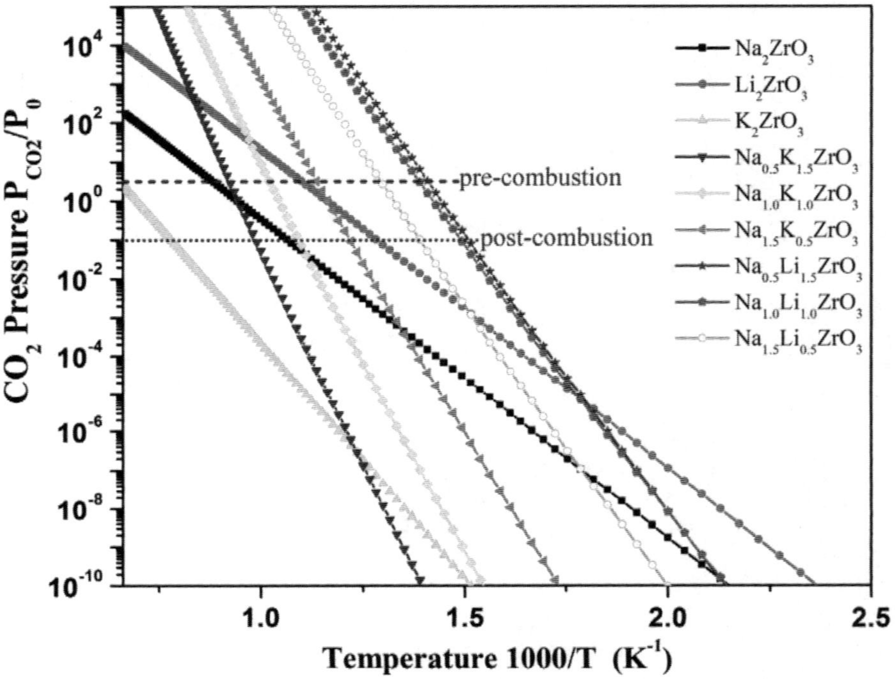

Fig. 6. The contour plotting of calculated chemical potential changes versus CO_2 pressures and temperatures of the reactions for $Na_{2-\alpha}M_\alpha ZrO_3$ (M=Li, K, α=0.5, 1.0, 1.5) capture CO_2. The typical CO_2 pressure for pre- and post-combustions are specified in blue lines.[33]

Acknowledgement

The author thanks Drs. D. C. Sorescu, D. Luebke, H. W. Pennline, B. Y. Li, H. Pfeiffer, J. K. Johnson, B. Zhang, K. Zhang, D. King, K. Parlinski for their kind help and collaborations.

References

[1] D. Aaron, and C. Tsouris, Separation Science and Technology **40**, 321 (2005).
[2] C. M. White, B. R. Strazisar, E. J. Granite, J. S. Hoffman, and H. W. Pennline, J Air Waste Manag Assoc **53**, 645 (2003).
[3] R. S. Haszeldine, Science **325**, 1647 (2009).
[4] B. Y. Li, Y. Duan, D. Luebke, and B. Morreale, Applied Energy **102**, 1439 (2013).
[5] M. R. Allen, D. J. Frame, C. Huntingford, C. D. Jones, J. A. Lowe, M. Meinshausen, and N. Meinshausen, Nature **458**, 1163 (2009).
[6] H. Pfeiffer, and P. Bosch, Chemistry of Materials **17**, 1704 (2005).
[7] E. Ochoa-Fernandez, H. K. Rusten, H. A. Jakobsen, M. Ronning, A. Holmen, and D. Chen, Catalysis Today **106**, 41 (2005).
[8] Y. Duan, D. Luebke, and H. W. Pennline, International Journal of Clean Coal and Energy **1**, 1 (2012).
[9] Y. Duan, and D. C. Sorescu, Phys Rev B **79**, 014301 (2009).
[10] Y. Duan, and D. C. Sorescu, J Chem Phys **133**, 074508 (2010).
[11] Y. Duan, in *Proceedings of 7th~13th Ann. Conf. on Carbon Capture, Sequestration & Utilization*Pittsburgh, 2008-2014).
[12] Y. Duan, and K. Parlinski, Phys Rev B **84**, 104113 (2011).
[13] Y. Duan, B. Zhang, D. C. Sorescu, and J. K. Johnson, J. Solid State Chem. **184**, 304 (2011).
[14] Y. Duan, D. R. Luebke, H. W. Pennline, B. Y. Li, M. J. Janik, and J. W. Halley, Journal of Physical Chemistry C **116**, 14461 (2012).
[15] Y. Duan, B. Zhang, D. C. Sorescu, J. K. Johnson, E. H. Majzoub, and D. R. Luebke, Journal of Physics-Condensed Matter **24**, 325501 (2012).
[16] B. Zhang, Y. Duan, and J. K. Johnson, J Chem Phys **136**, 064516 (2012).
[17] Y. Duan, Phys. Rev. B **77**, 045332 (2008).
[18] Y. Duan, J Renew Sustain Ener **3**, 013102 (2011).
[19] Y. Duan, J Renew Sustain Ener **4**, 013109 (2012).
[20] M. W. J. Chase, J. Phys. Chem. Ref. Data, **Monograph 9**, 1 (1998).
[21] Y. Duan, Physical Chemistry Chemical Physics **15**, 9752 (2013).
[22] Y. Duan, H. Pfeiffer, B. Li, I. C. Romero-Ibarra, D. C. Sorescu, D. R. Luebke, and J. W. Halley, Physical Chemistry Chemical Physics **15**, 13538 (2013).
[23] Y. Duan, H. Pfeiffer, B. Y. Li, I. C. Romero-Ibarra, D. C. Sorescu, D. R. Luebke, and J. W. Halley, Physical Chemistry Chemical Physics **15**, 13538 (2013).
[24] Y. Duan, J Renew Sustain Ener **4**, 013109 (2012).
[25] K. L. Zhang, X. H. S. Li, Y. Duan, D. L. King, P. Singh, and L. Y. Li, International Journal of Greenhouse Gas Control **12**, 351 (2013).
[26] K. L. Zhang, X. H. S. Li, W. Z. Li, A. Rohatgi, Y. Duan, P. Singh, L. Y. Li, and D. L. King, Advanced Materials Interfaces **1**, 1400030 (2014).
[27] J. L. Chi, L. F. Zhao, B. Wang, Z. Li, Y. H. Xiao, and Y. Duan, International Journal of Hydrogen Energy **39**, 6479 (2014).
[28] Y. Duan, K. Zhang, X. S. Li, D. L. King, B. Li, L. Zhao, and Y. Xiao, Aerosol and Air Quality Research **14**, 470 (2014).
[29] Y. Duan, ScienceJet **3**, 56 (2014).
[30] M. Y. Veliz-Enriquez, G. Gonzalez, and H. Pfeiffer, J Solid State Chem **180**, 2485 (2007).
[31] H. Pfeiffer, E. Lima, and P. Bosch, Chemistry of Materials **18**, 2642 (2006).
[32] H. Pfeiffer, C. Vazquez, V. H. Lara, and P. Bosch, Chemistry of Materials **19**, 922 (2007).
[33] Y. Duan, J. Lekse, X. F. Wang, B. Y. Li, B. Alcantar-Vazquez, H. Pfeiffer, and J. W. Halley, Physical Review Applied, accepted (2015).

399080 Zero Ammonia Gas Emmission Strategy @ Ammonia Urea Complex

Abstract:

Engro Fertilizers limited is 2nd largest Urea Producer in Pakistan. Engro is operating **02** Ammonia, **03** Urea and **03** Utilities plants near a rich natural gas facility in district Ghokti with total capacity of **2.1** Million tons per year. In 2014 Engro achieved land mark **DuPont 4.3** ranking in process safety management. High standard of process safety has been achieved through top management commitment and developing a marvelous system of integrated safety management, now Engro is striving for world class safety and Environment performer. Engro commissioned new Ammonia Urea complex in **2010** based on Snamprogetti technology having **2194 MeT/day** Liquid Ammonia and **3835 MeT/day** Urea capacity.

Engro Fertilizers HSE policy is true depiction of its top management commitment towards Process Safety and Environment Management. One clause of policy says *"Continuously improve environmental performance to achieve sustainable development."* In this regard state of the art technology, flare system was procured for new Ammonia Urea complex to ensure zero ammonia gas effluent during normal operation, start up, shutdown and process upset. This initiative eradicated the all possible risks related to ammonia release in environment. Flare system protected workforce at plant, surrounding community and trespassers. It has been designed to burn continuous ammonia effluent as well as to cater abrupt ammonia release in case of process upset. Since commissioning of new Ammonia Urea plant, environment complains have been drastically reduced and one element of environment pollution have been eradicated. Moreover detail PHA studies were conducted to analyze the relevant hazards and their consequences. After commissioning, a key in house design modification was conducted to avoid localized Ammonical water release at Urea plant, which not only avoided ammonia release from Urea plant permissive but also resulted in ammonia recovery through close header, recovered ammonia recycled to Urea process which enhanced Urea yield.

399820 Importance of Process and Flaring Data and Its Analysis for the Management of Abnormal Situations: An Ethylene Process Case Study

Abstract:

Upset emissions occur during plant startup, shutdown, maintenance, malfunction and flaring incidents. A wide range of these Upsets cannot be managed by the standalone control systems even the advanced ones and need the interfere of plant personnel to correct the causes of this abnormal situation. The technologies needed to help plant personnel to control and prevent abnormal situations are gathered under abnormal situation management. These situations have significant impact on the process industry economy, safety and environment. Moreover, flaring, which is a key but negative method for managing abnormal situations in petrochemical plants, emits huge amounts of greenhouse gases (GHG) and hazardous volatile organic compounds (VOC_S), causing locally transient air pollution and negative community impacts. In addition, flaring is causing huge loss in raw material and energy which could be used to generate more desired products for the industry. Thus, flare minimization during normal and abnormal operational situations has great benefits to environmental, societal, and industrial sustainability. It is not possible to quantify the impacts without understanding the properties and magnitude of these upsets. The analysis of the properties and magnitude of the upset emissions needs extensive amount of historical data. In this paper, an ethylene base case study along with its historical process and flaring incidents data are used to identify underlying causes and indicators that lead to process upsets, i.e. abnormal situations. The paper will also show potential technique(s) that are recommended to mitigate the incidents and reduce/eliminate its root causes. Moreover, a systematic approach to collect and analyze historical flaring data based on current industrial practices is presented. Finally, the challenges that are faced by environmental agencies in terms of data utilization and documentation are highlighted with some recommended improvements and suggestions.

401586 Aspen Plus Simulation of Steam-Biogas Reforming and Comparison with Experimental Data Obtained over a Metal-Foam-Coated Pd-Rh Catalyst

Abstract:

The steam biogas reforming (SBR) process was studied using the Aspen Plus process simulation tool. Model biogas (60% CH_4 + 40% CO_2) mixed with steam at S/C (steam to methane) ratios of 1.25, 1.50 and 1.75 was used as the feedstock. SBR equilibrium data were obtained over a temperature range of 923 to 1123 K at 1 atm pressure through Gibbs free energy minimization. The equilibrium data were compared with experimental data obtained over a [Pd(7)-Rh(1)/($CeZrO_2$•Al_2O_3)]/metal foam catalyst at a GHSV of 20,000 h^{-1}. At 923 K, CH_4 conversion over the catalyst was approximately 15% lower than the equilibrium conversion, but the difference was attenuated with increasing temperature and was completely removed around 1123 K. The CO/(CH_4 + CO_2) yields over the catalyst were lower than the equilibrium data in the entire temperature range due to the lower CH_4 as well as CO_2 conversions. The H_2/CH_4 yields over the catalyst were lower initially, but exceeded the equilibrium data at the temperature around 1050 K due to water gas shift (WGS) reaction being favored over the catalyst. The WGS reaction was responsible for the lower CO_2 conversion. The process thermal efficiencies were calculated through the simulation and experimental data and the results were comparable. The comparisons of the experimental data with the Aspen Plus simulation results will help to enable the SBR process optimization with a focus on improving the net thermal efficiency and product syngas yield.

References

1. Wheeler C, Jhalani A, Klein EJ, Tummala S, Schmidt LD, The water–gas-shift reaction at short contact times. *J catal.* 2004;223(1):191-199.

2. Roy PS, Raju ASK, Kim K, Influence of S/C ratio and temperature on steam reforming of model biogas over a metal-foam-coated Pd–Rh/($CeZrO_2$–Al_2O_3) catalyst. *Fuel*. 2015;139:314-320.

3. Roy PS, Park N-K, Kim K, Metal foam-supported Pd–Rh catalyst for steam methane reforming and its application to SOFC fuel processing. *Int J Hydrogen Energy*. 2014;39(9):4299-4310.

400201 A Production Possibility Frontier of a Sustainable Coal, Wind, and SMALL Modular Reactor Hybrid Energy System

Abstract:

A PRODUCTION POSSIBILITY FRONTIER OF A SUSTAINABLE COAL, WIND, AND SMALL MODULAR REACTOR HYBRID ENERGY SYSTEM

Kyle L. Buchheit and Joseph D. Smith, Ph.D.

Missouri University of Science and Technology, Rolla Missouri

ABSTRACT

The availability of cheap electricity is one of the biggest factors for improving quality of life. With the debate on the effects of carbon dioxide emissions continuing, several countries have either implemented or are considering the reduction of emissions through various economic means. The inclusion of a monetary penalty on carbon emissions would increase the prices of electricity produced by carbon-based sources. The push for large-scale renewable sources of energy has met problems with regards to energy storage and availability. The proposed coal, wind and nuclear hybrid energy system would combine a renewable energy source, wind, with traditional and stable energy sources, coal and nuclear, to create an integrated and sustainable system. Instead of a light water reactor, a next generation small modular nuclear reactor will be evaluated. The coal system will use a pressurized circulating fluidized bed system, which can not only take coal, but also biomass as a carbon feedstock. This system also employs a high temperature steam co-electrolysis unit for the utilization of carbon dioxide emissions for the production of synthetic gas which can be used in the production of transportation fuels or chemicals.

An existing rigorous dynamic process model will be used to simulate the potential output of the system based on real world dynamic data. System inputs will include a full year of dynamic wind speeds for variable power generation and simulated electrical grid demand. These inputs will vary the amount of power available for synthetic gas production, and thus yield a theoretical production frontier over a year of operation. This information will be used to determine overall process economics based on sale of synthetic gas and levelized cost of electricity.

15AIChE
Spring Meeting, Austin, TX

Thermochemical Conversion of H_2O and CO_2 into Solar Fuels via Metal Oxide Based Redox Reaction

(Principal Author)
Shahd Gharbia
Department of Chemical Engineering
College of Engineering
Qatar University, Doha, Qatar
sg1000354@student.qu.edu.qa

(Other Authors)
Rahul Bhosale
Department of Chemical Engineering
College of Engineering
Qatar University, Doha, Qatar
rahul.bhosale@qu.edu.qa

Anand Kumar
Department of Chemical Engineering
College of Engineering
Qatar University, Doha, Qatar
akumar@qu.edu.qa

Fares A. AlMomani
Department of Chemical Engineering
College of Engineering
Qatar University, Doha, Qatar
falmomani@qu.edu.qa

Dareen Dardor
Department of Chemical Engineering
College of Engineering
Qatar University, Doha, Qatar
dd1202223@student.qu.edu.qa

Moustafa Hussein Ali
Department of Chemical Engineering
College of Engineering
Qatar University, Doha, Qatar
ma1203561@student.qu.edu.qa

Jamila Folady
Department of Chemical Engineering
College of Engineering
Qatar University, Doha, Qatar
jf1003686@student.qu.edu.qa

Shiva Yousefi
Department of Chemical Engineering
College of Engineering
Qatar University, Doha, Qatar
sy1106491@student.qu.edu.qa

Mehak Jilani
Department of Chemical Engineering
College of Engineering
Qatar University, Doha, Qatar
mj1104333@student.qu.edu.qa

Mashail Alfakih
Department of Chemical Engineering
College of Engineering
Qatar University, Doha, Qatar
ma090163@student.qu.edu.qa

Prajakta Angre
Department of Chemical Engineering
MGM College of Engineering & Technology
Mumbai University, Navi Mumbai
prajaktaangre752007@gmail.com

Prepared for Presentation at
American Institute of Chemical Engineers
2015 Spring Meeting
11th Global Congress on Process Safety
Austin, Texas
April 27-30, 2015

UNPUBLISHED

Thermochemical Conversion of H_2O and CO_2 into Solar Fuels via Metal Oxide Based Redox Reaction

Shahd Gharbia
Department of Chemical Engineering
College of Engineering
Qatar University, Doha, Qatar
sg1000354@student.qu.edu.qa

Rahul R. Bhosale[1], Anand Kumar[1], Fares A. AlMomani[1], Dareen Dardor[1], Moustafa Hussein Ali[1], Jamila Folady[1], Shiva Yousefi[1], Mehak Jilani[1], Mashail Alfakih[1], Prajakta Angre[2]

[1]Department of Chemical Engineering
College of Engineering
Qatar University, Doha, Qatar

[2]Department of Chemical Engineering
MGM College of Engineering & Technology
Mumbai University, Navi Mumbai, India

Keywords: Solar Energy, Chemical Equilibrium Thermodynamics, Zinc Oxide, Alternative Fuels, Thermochemical Redox Reactions, Second Law Analysis

Abstract

This paper reports the thermodynamic analysis of the production of solar syngas (a mixture of H_2 and CO with molar ratio equal to 2:1) via ZnO/Zn based thermochemical H_2O and/or CO_2 splitting redox reactions. Solar syngas production from CO_2 and H_2O that is considered in a two-step thermochemical cycle via Zn/ZnO redox reactions, encompassing: 1) the ZnO thermolysis to Zn and O_2 using concentrated solar radiation as the source of process heat, and 2) Zn reacting with mixtures of H_2O and CO_2 yielding high-quality syngas and ZnO. The ZnO is recycled to the first, solar step. Syngas is further processed to liquid hydrocarbon fuels via FischereTropsch or other catalytic processes. Second-law thermodynamic analysis is applied to determine the cycle efficiencies attainable with and without heat recuperation for varying the thermal reduction temperature (T_R) in the range 1700-1950 K. Furthermore, the effect of various T_R on molar flow rate of inert Ar has been presented. Considered is the energy penalty of using Ar dilution in the solar step below 2235 K for shifting the equilibrium to favor Zn production. Thermodynamic analysis has been performed and the analysis shows that solar syngas production via the Zinc Oxide redox cycle is a promising approach for CO_2/H_2O conversion into alternative fuels.

Note: Do not add page numbers. Do not refer to page numbers when referencing different portions of the paper

1. Introduction

The world has been obsessed with the shortage of fossil fuels for decades and the main challenge for most countries is securing fuel supplies. Apart from the fuel demands, the problem of CO_2 emission is another global issue concerning all societies. CO_2 is the main reason causing the greenhouse effect, which leads to global warming and other consequences such as glaciers melting, sea level rise, and extreme climates, etc. The CO_2 emission from fuel combustion was 31.6 Gt (billion metric tons) in 2011 [1-2]. As a result, it is highly essential to invest towards renewable energy technologies such as solar energy (storage, conversion and utilization). Through solar fuel processes, efficient chemical storage of intermittent sunlight can be realized by using concentrated solar energy as the high temperature heat source to drive highly endothermic processes for producing solar fuels [3,4]. The production of CO and H_2 constitutes synthesis gas (syngas), the precursor to liquid hydrocarbon fuels via Fischere Tropsch or other catalytic processes. Syngas produced with concentrated solar energy from H_2O and CO_2, the latter eventually captured from air [5-8], signifies a promising path toward sustainable energy utilization. [9] Moreover, due to the challenges of demands on alternative fuels and CO_2 emission, the conversion of CO_2 has become a hot spot.

Among various methods, two-step conversion of CO_2 with Zinc Oxide appears to be a promising way. [10]. The present work considers the approaches for syngas production through solar fuels via the Zinc Oxide redox cycle. In order to evaluate the capability of the system for exporting work, the system was analyzed on the basis of the Second Law of Thermodynamics and the reaction mechanism of Zinc Oxide taking into consideration that for high quality syngas, the H_2:CO ratio is in the order of 2:1. [12] In previous investigations, various MO based redox systems were investigated for the solar thermochemical H_2O and/or CO_2 splitting such as ZnO/Zn [11], SnO_2/SnO [12-16], Fe_3O_4/FeO [17-18], mixed ferrites [19-34], and ceria based oxides [35-40].

In this investigation, thermodynamic analysis of the ZnO/Zn based solar thermochemical H_2O and CO_2 splitting process was investigated in two sections with the help of commercial thermodynamic HSC Chemistry software and databases [37]. In first section, chemical thermodynamic equilibrium of the ZnO/Zn based solar thermochemical H_2O and CO_2 splitting process was studied by investigating the effect of various operating parameters on the thermal reduction and H_2O/CO_2 splitting temperature. In second section, the second law thermodynamic analysis of this process was performed. The solar absorption efficiency of the solar reactor ($\eta_{absorption}$), solar energy input to the solar reactor (Q_{solar}), net energy absorbed in the solar reactor ($Q_{reactor-net}$), rediation heat losses from the solar reactor ($Q_{re-radiation}$), rate of heat rejected to the surrounding from the quench unit (Q_{quench}), and from the Zn oxidizer (Q_{Zn-Oxd}) , and solar to chemical energy conversion efficiency ($\eta_{solar-to-chemical}$) were determined by performing the thermodynamic modeling and are presented in this paper. A typical ZnO/Zn based two-step solar thermochemical H_2O and CO_2 splitting process is presented in Figure 1.

2. Chemical Thermodynamic Equilibrium

2.1 Thermal Reduction of Zinc Oxide

Reaction (1) represents the thermal reduction of ZnO into Zn.

$$3ZnO(s) + n\,Ar(g) \rightarrow 3Zn(g) + 1.5\,O_2(g) + n\,Ar(g) \qquad (1)$$

According to reaction (1), thermal reduction of ZnO into Zn can be summarized as the conversion of solid reactants into gaseous products. Effect of moles of inert gas (*Ar*) used during the thermal reduction of ZnO on the thermal reduction temperature was examined and presented in Figure 2. The data reported shows that as the thermal reduction temperature of ZnO decreases as the quantity of *Ar* in the feed increases. For instance, if we used 680 mol/sec of Ar flow rate, the 100% thermal reduction can be achieved at 1700 K. In contrast to this, higher thermal reduction temperature (1950 K) is needed if 45 mol/sec of Ar flow rate is applied.

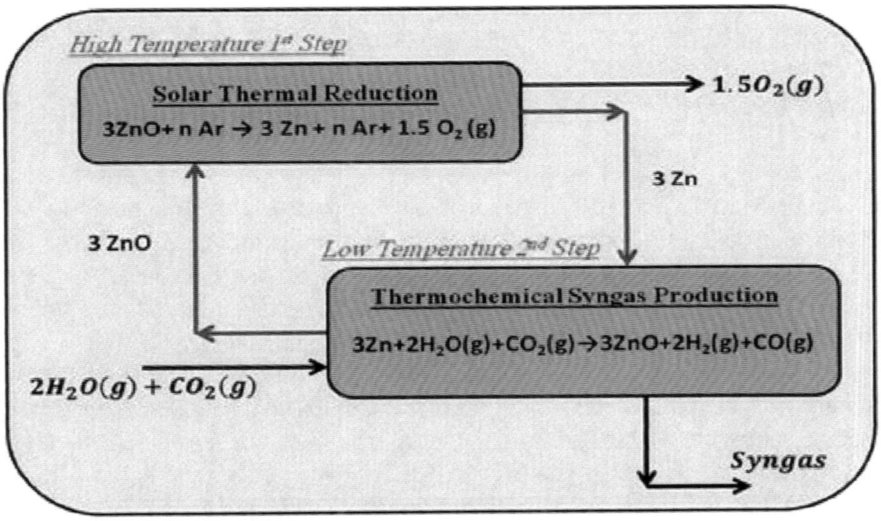

Figure 1: Zinc Oxide based solar thermochemical syngas production process.

2.2 Syngas Production

Reaction (2) represents the production of syngas (H_2 and CO with molar ratio equal to 2) by oxidation of Zn via combined H_2O and CO_2 splitting reaction.

$$3ZnO(s) + 2H_2O + CO_2(g) \rightarrow 3ZnO(s) + 2H_2(g) + CO(g) \qquad (2)$$

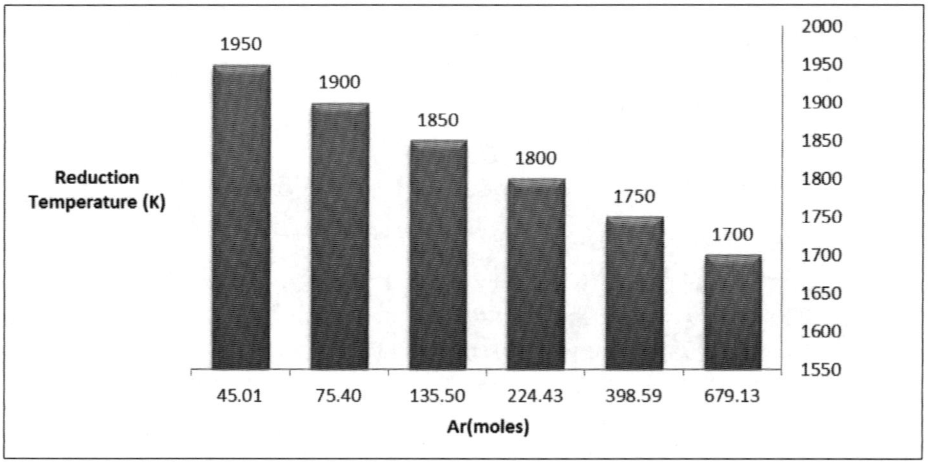

Figure 2: Effect of Ar flowrate on thermal reduction temperature of ZnO.

3. Second Law Analysis

3.1 Process Flow Diagram for ZnO/Zn Based Syngas Production Process

This section reports the second law analysis of the ZnO/Zn based two-step solar thermochemical H_2O and CO_2 splitting process. A process flow diagram for this solar syngas production process is shown in Figure 3. The continuous ZnO/Zn based syngas production process is consists of a solar reactor, a quench unit, a Zn oxidizer, gas separator and theoretical $H_2/CO/O_2$ ideal fuel cell. The molar flow rates of reactants are shown in Figure 3. It is a steady state process and all the reactions are carried out at 1 atm. A few assumptions have been included: a) the solar reactor is assumed to be a perfectly insulated, blackbody absorber, b) no convective/conductive heat losses, c) effective emissivity and absorptivity equal to unity, d) kinetic and potential energies are neglected, e) all reactions reach 100% completion. Thermodynamic properties were extracted from HSC software and databases and the analysis is normalized to the ZnO molar flow rate (3 mol/sec) entering the solar reactor.

3.2 Solar Energy Absorption Efficiency $(\eta_{absorption})$

The solar energy absorption efficiency $(\eta_{absorption})$ is defined as the net rate at which energy is being absorbed by the solar reactor divided by the solar power input through the aperture of the solar reactor. The $\eta_{absorption}$ can be calculated as:

$$\eta_{absorption} = \frac{Q_{reactor-net}}{Q_{solar}} = 1 - \left(\frac{\sigma T_H^4}{IC}\right) \quad (3)$$

Where, σ = Stefan-Boltzmann constant, I = normal beam insolation $1(kW/m^2)$, C= solar flux concentration ratio of the solar concentrating system (1000 suns), $Q_{reactor-net}$= net energy absorbed in the solar reactor, Q_{solar} = solar energy input. For SnO_2/SnO based syngas production process, at T_H= 1900K, the $\eta_{absorption}$ was observed to be equal to 26.10%.

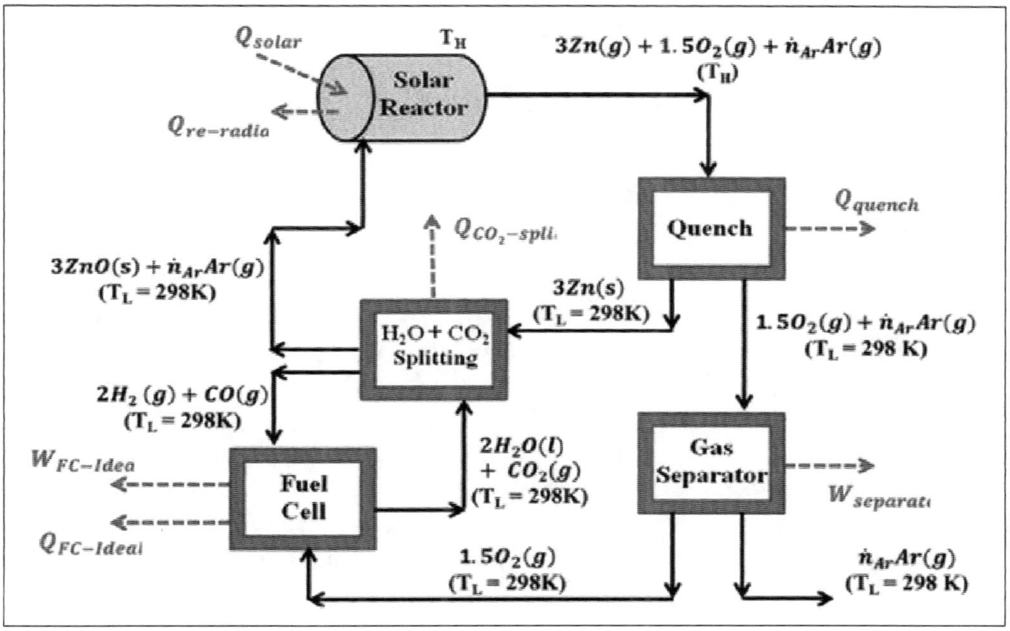

Figure 3: The flow diagram of a two-step solar thermochemical ZnO/Zn based syngas production process.

3.3 Net Solar Energy Absorbed for The Operation of SO-SS Cycle

According to Figure 3, 3 moles of ZnO(s) enters the solar reactor at T_R= 298K and then heated upto T_H= 1900K with the help of solar energy entering in the solar reactor from the solar concentrating system and undergoes 100% thermal reduction. The net solar energy absorbed by the solar reactor can be calculated as:

$$Q_{reactor-net} = Q_{ZnO-heating} + Q_{ZnO-reduction} + Q_{Ar-heating} \tag{4}$$

Thermodynamic calculations reveal that the 285.45 kW of heat is required for the heating of ZnO from 298K upto 1900K $\left(Q_{ZnO-heating}\right)$. Furthermore, the amount of solar energy utilized for the thermal reduction of ZnO at 1900K [according to reaction (1)] equals to 1347.24 kW. Also 2510.73 kW is required to heat Ar from 298K upto 1900K. Therefore, the $Q_{reactor-net}$ for the solar reactor used for this process is equal to 4143.45 kW.

3.4 Solar Energy Input To The Solar Reactor (Q_{solar})

Based on the calculated $\eta_{absorption}$ and $Q_{reactor-net}$, the amount of solar energy input required to operate the solar reactor (Q_{solar}) dealing with the ZnO/Zn based syngas production process can be determined by equation (5) and observed to be equal to 15875 kW.

$$Q_{solar} = \frac{Q_{reactor-net}}{\eta_{absorption}} \tag{5}$$

3.5 Radiation Heat Losses from The Solar Reactor

Radiation heat losses from the solar reactor conducting the thermal reduction of ZnO can be calculated by equation (6):

$$Q_{re-radiation} = Q_{Solar} - Q_{reactor-net} \tag{6}$$

For the solar reactor conducting the ZnO/Zn based solar thermochemical H_2O and/or CO_2 splitting process, the $Q_{re-radiation}$equals to 11731.55 kW.

3.6 Rate of Heat Rejected To The Surrounding Due To Quenching

After performing thermal reduction of ZnO, gaseous products Zn, O_2 and Ar exiting the solar reactor are rapidly cooled down from thermal reduction temperature (1900K) to ambient temperature (298K). The purpose of this quenching step is to avoid the recombination of the gaseous products. The rates of heat rejected by the quenching unit (Q_{quench}) during this step is calculated by the following equation (7) and observed to be 3091.95 kW.

$$Q_{quench} = -\dot{n}\Delta H|_{3Zn(g)+1.5O_2(g)+\ 75\mathrm{Ar(g)}@1900K \rightarrow 3Zn(s)+1.5O_2(g)+75\mathrm{Ar(g)}@298K} \tag{7}$$

3.7 Rate of Heat Rejected To The Surrounding from The Zn Oxidizer (Q_{Zn-oxd})

Condensed Zn from the quench unit is transferred to the Zn oxidizer and reacted with the combined stream of H_2O and CO_2 to form syngas and ZnO(s). It is noteworthy that the heat released from the exothermic H_2O and CO_2 splitting reaction is sufficient enough to overcome the sensible and latent heats and raise the temperature from 298K to the required splitting temperature. For this step, complete conversion of Zn to ZnO is assumed and the rate of heat rejected to the surrounding from the Zn oxidizer $(Q_{Zn-oxidizer})$ is estimated to be 197.03 kW [equation (8)].

$$Q_{Zn-oxd} = -\dot{n}\Delta H|_{3Zn(s)+2H_2O(l)+CO_2(g) \rightarrow 3ZnO(s)+2H_2(g)+CO(g)@298K} \tag{8}$$

3.8 Rate of Theoretical Work Performed And Heat Rejected By The Ideal Fuel Cell

The ZnO/Zn based syngas production cycle is completed with the addition of an ideal fuel cell operated in presence of $H_2/CO/O_2$. The fuel cell is added to the cycle to extract the maximum work of the net syngas produced by this process. The rates of theoretical work performed $(W_{FC-Ideal})$ and the heat rejected by the fuel cell $(Q_{FC-Ideal})$ can be calculated by using following equations (10) and (11). For This calculation the efficiency of the fuel cell (η_{FC}) is assumed to be 100%.

$$W_{FC-Ideal} = -(\eta_{FC}) \times \dot{n}\ \Delta G|_{2H_2(g)+CO(g)+1.5O_2(g) \rightarrow 2H_2O(l)+CO_2(g)@298K} \tag{9}$$

$$Q_{FC-Ideal} = -(T_R) \times \dot{n}\ \Delta S|_{2H_2(g)+CO(g)+1.5O_2(g) \rightarrow 2H_2O(l)+CO_2(g)@298K} \tag{10}$$

For the ZnO/Zn based solar thermochemical syngas production process described in Figure 3, the $W_{FC-Ideal}$ and $Q_{FC-Ideal}$ are equal to 731.65 kW and 122.865 kW, respectively.

3.9 Work Done By The Separator

The work that is required to separate Ar from O_2 (gas) can be calculated from equation (11) and observed to be equal to 75.40 kW

$$W_{Seperator} = \dot{n}RT_{amb}\left(0.5\ln\frac{1}{y_{O_2}} + \frac{n_{Ar}}{n_{ZnO}} \times \ln\frac{1}{y_{Ar}}\right) \quad (11)$$

3.10 Solar To Chemical Energy Conversion Efficiency ($\eta_{solar-to-chemical}$)

The solar to chemical energy conversion efficiency ($\eta_{solar-to-chemical}$) is defined as the portion of solar energy that is converted into the chemical energy by the Gibbs free energy of the products (i.e. maximum possible work that can be extracted from the products).

$$_{solar-to-chemical} = \frac{W_{FC-Ideal}}{Q_{solar}} \quad (12)$$

The $\eta_{solar-to-chemical}$ of the ZnO/Zn solar thermochemical syngas production process is calculated using equation (12) and is equal to 4.57%.

4. Conclusion

Two-step solar thermochemical cycles for syngas production via ZnO/Zn based H_2O and CO_2 splitting reactions have been thermodynamically examined. Thermodynamic equilibrium calculations performed at 1 atm pressure revealed that the exact number of moles of Ar gas in the feed can be significantly decreased is by increasing temperature for the thermal reduction of ZnO into Zn. The second law thermodynamic analysis of the solar reactor conducting the thermochemical ZnO/Zn based H_2O and CO_2 splitting process at 1900 K and at a solar concentration of 1000 suns indicate $_{absorption}$ = 26.10%, Q_{solar}= 15875 kW, $Q_{reactor-net}$ = 4143.45 kW, $Q_{re-radiation}$ = 11731.55 kW, Q_{quench}= 3091.95 kW, Q_{Zn-oxd} = 197.03 kW, and $_{solar-to-chemical}$ = 4.57%, respectively.

5. Acknowledgments

The authors gratefully acknowledge the financial support provided by the Qatar University Internal Grants QUUG-CENG-CHE-13/14-4 and QUUG-CENG-CHE-14\15-10.

References

(1) Olivier JGJ, Janssens-Maenhout G, A.H.W. Peters J. Trends in global CO_2 emissions 2012; 2012.

(2) Fakhrai, R., Saadatfar, B., & Wei, B., The thermodynamic analysis of two-step conversions of CO_2/H_2O for syngas production by ceria. International Journal of Hydrogen Energy, 39 (2014) 12353.

(3) Steinfeld A, Weimer AW. Thermochemical production of fuels with concentrated solar energy. Optics Express, 18 (2010) A100.

(4) Steinfeld A. Solar thermochemical production of hydrogen: a review. Solar Energy, 78 (2005) 603.

(5) Nikulshina V, Galvez ME, Steinfeld A. Kinetic analysis of the carbonation reactions for the capture of CO_2 from air via the $Ca(OH)_2$–$CaCO_3$–CaO solar thermochemical cycle. Chemical Engineering Journal, 129 (2007) 75.

(6) Nikulshina V, Gebald C, Steinfeld A. CO_2 capture from atmospheric air via consecutive CaO-carbonation and $CaCO_3$-calcination cycles in a fluidized-bed solar reactor. Chemical Engineering Journal, 146 (2009) 244.

(7) Nikulshina V, Hirsch D, Mazzotti M, Steinfeld A. CO_2 capture from air and co-production of H_2 via the $Ca(OH)_2$–$CaCO_3$ cycle using concentrated solar power-thermodynamic analysis. Energy, 31 (2006) 1715.

(8) Nikulshina V, Steinfeld A. CO_2 capture from air via CaO carbonation using a solar-driven fluidized bed reactor effect of temperature and water vapor concentration. Chemical Engineering Journal, 155 (2009) 867.

(9) Davidson, J., & Krenzke, P. Thermodynamic Analysis of Syngas Production via the Solar Thermochemical Cerium Oxide Redox Cycle with Methane-Driven Reduction. Energy & Fuels, 28 (2014) 4088.

(10) Bhosale, R., Yousefi, S., Folady, J., Jilani, M., Dardor, D., Gharbia, S., Kumar, A., van den Broeke, LJP., Alxneit, I., Shende, I. Thermodynamic analysis of solar fuel production via thermochemical H_2O and/or CO_2 splitting using tin oxide based redox reactions. Proceedings of 4th GPC Symposium (2014) (In Press).

(11) Loutzenhiser, P., Steinfeld, A. Solar syngas production from CO_2 and H_2O in a two-Step Thermochemical Cycle via Zn/ZnO Redox Reactions: Thermodynamic cycle Analysis. International Journal of Hydrogen Energy, 36 (2011) 12141.

(12) Stamatiou, A., Loutzenhiser, P., Steinfeld, A. Solar Syngas Production from H_2O and CO_2 via two-Step Thermochemical Cycles based on Zn/ZnO and FeO/Fe_3O_4 Redox Reactions: Kinetic Analysis. Energy & Fuels, 24 (2010) 2716.

(13) Charvin, P., Abanades, S., Lemont, F., Flamant, G. Experimental study of SnO_2/SnO/Sn thermochemical systems for solar production of hydrogen. AIChE Journal, 54 (2008) 2759.

(14) Abanades, S. CO_2 and H_2O reduction by solar thermochemical looping using SnO_2/SnO redox reactions: Thermogravimetric analysis. International Journal of Hydrogen Energy, 37 (2012) 8223.

(15) Chambon, M., Abanades, S., Flamant, G. Kinetic investigation of hydrogen generation from hydrolysis of SnO and Zn solar nanopowders. International Journal of Hydrogen Energy, 34 (2009) 5326.

(16) Vishnevetsky, I., Epstein, M., J. Tin as possible candidate for solar thermochemical redox process for hydrogen production. Journal of Solar Energy Engineering, 131 (2009) 021007-1.

(17) Abanades, S., Villafan-Vidales, H. CO_2 and H_2O conversion to solar fuels via two-step solar thermochemical looping using iron oxide redox pair. Chemical Engineering Journal 175 (2011) 368.

(18) Gokon, N., Murayama, H., Nagasaki, A., Kodama, T. Thermochemical two-step water splitting cycles by monoclinic ZrO_2 supported $NiFe_2O_4$ and Fe_3O_4 powders and ceramic foam. Solar Energy, 83 (2009) 527.

(19) Scheffe, J., Li, J., Weimer, A. A spinel ferrite/hercynite water-splitting redox cycle International Journal of Hydrogen Energy, 35 (2010) 3333.

(20) Roeb, M., Gathmann, N., Neises, M., Sattler, C., Pitz-Paal, R. Thermodynamic analysis of two-step solar water splitting with mixed iron oxides. International Journal of Energy Research, 33 (2009) 893.

(21) Lorentzou, S., Agrafiotis, A., Konstandopoulos, A., Aerosol spray pyrolysis synthesis of water-splitting ferrites for solar hydrogen production. Granular Matter, 10 (2008) 113.

(22) Bhosale, R., Shende, R., Puszynski, J., H_2 Generation from Thermochemical Water-Splitting Using Sol-Gel Derived Ni-Ferrite. Journal of Energy and Power Engineering, 4 (2010) 27.

(23) Bhosale, R., Shende, R., Puszynski, J., H_2 Generation from Thermochemical Water-Splitting Using Sol-Gel Synthesized Zn/Sn/Mn-Doped Ni-Ferrite. International Review of Chemical Engineering 2 (2010) 852.

(24) Bhosale, R., Khadka, R., Shende, R., Puszynski, J., H_2 Generation from Two-Step Thermochemical Water-Splitting Reaction Using Sol-Gel Derived SnxFeyOz. Journal of Renewable and Sustainable Energy, 3 (2011) 063104-1.

(25) Bhosale, R., Shende, R., Puszynski, J., Thermochemical Water-Splitting for H2 Generation Using Sol-Gel Derived Mn-Ferrite In A Packed Bed Reactor. International Journal of Hydrogen Energy, 37 (2012) 2924.

(26) Bhosale, R., Shende, R., Puszynski, J., Sol-Gel Derived NiFe2O4 Modified With ZrO2 for Hydrogen Generation from Solar Thermochemical Water-Splitting Reaction. Mater. Res. Soc. Symp. Proc., 1387 (2012).

(27) Bhosale, R., Alxneit, I, van den Broeke, Leo L. P., Kumar, A., Jilani, M., Gharbia, S., Folady, J., Dardor, D., Solar Fuel Production via Non-Stoichiometric $Ce_xZr_yHf_zO_{2-\delta}$ Based Two-Step Thermochemical Redox Cycle. Mater. Res. Soc. Symp. Proc., 1675, mrss14-1675-rr06-10 (2014).

(28) Bhosale, R., Shende, R., Opoku, M. K. Puszynski, J., H_2 Generation from Water-Splitting Using Novel Ferrite Foam Materials. AIChE Annual Meeting Proc. (2009).

(29) Bhosale, R., Shende, R., Puszynski, J., H2 Generation from Thermochemical Water-Splitting Using Sol-Gel Synthesized Ferrites. AIChE Annual Meeting Proc. (2010).

(30) Pasala, X., Bhosale, R., Amar, V. S., Puszynski, J., Shende, R. Thermally Stabilized Ferrite Nanoparticles for Hydrogen Generation from Thermochemical Water-Splitting Reaction. AIChE Annual Meeting Proc. (2012).

(31) Shende, R., Puszynski, J., Opoku, M., Bhosale, R., Synthesis of Nano-Sized Ferrite Materials for H2 Generation from High Temperature Water-Splitting Reaction. NSTI-Nanotech Conference & Expo. Proc. ISBN 978-1-4398-1782-7, 1 (2009) 201.

(32) Bhosale, R., Shende, R., Puszynski, J. Synthesis of Novel Ferrite Foam Materials for Water-Splitting Application. NSTI-Nanotech Conference & Expo. Proc. ISBN 978-1-4398-3401-5, 1 (2010) 368.

(33) Bhosale, R., Shende, R., Puszynski, J. Synthesis of Nano-Sized Ferrite Materials for H2 Generation from High Temperature Water-Splitting Reaction. NSTI-Nanotech Conference & Expo. Proc. ISBN 978-1-4398-7142-3, 1 (2011) 237.

(34) Bhosale, R., Pasala, X., Yelakanti, S., Puszynski, J., Shende, R. Solar Thermochemical Water-Splitting for H2 Generation Using Sol-Gel Derived Ferrite Nanomaterials. NSTI-Nanotech Conference & Expo. Proc. ISBN 978-1-4665-6276-9, 3 (2012) 555.

(35) Abanades, S., Flamant, G. Thermochemical hydrogen production from a two-step solar-driven water-splitting cycle based on cerium oxides. Solar Energy, 80 (2006) 1611.

(36) Chueh, W. C., Falter, C., Abbott, M., Scipio, D., Furler, P., Haile, S., Steinfeld, A. High-Flux Solar-Driven Thermochemical Dissociation of CO_2 and H_2O using Nonstoichiometric Ceria. Science, 330 (2010) 1797.

(37) Kaneko, H., Miura, T., Ishihara, T., Taku, S., Yokoyama, T., Nakajima, H., Tamaura, Y. Cerium ion redox system in CeO2-xFe2O3 solid solution at high temperatures (1273-1673K) in the two-step water splitting reaction for solar H_2 generation. Energy, 43 (2008) 3153.

(38) Meng, Q., Lee, C., Ishihara, T., Kaneko, H., Tamaura, Y. Reactivity of CeO2-based ceramics for solar hydrogen production via a two-step water-splitting cycle with concentrated solar energy. International Journal of Hydrogen Energy, 36 (2011) 13435.

(39) Le Gal, A., Abanades, S. Catalytic investigation of ceria-zirconia solid solutions for solar hydrogen production. International Journal of Hydrogen Energy, 36 (2011) 4739.

(40) Abanades, S., Gal, A. L. CO_2 Splitting by Thermo-Chemical Looping Based on $Zr_xCe_{1-x}O_2$ Oxygen Carriers for Synthetic Fuel Generation. Fuel, 102 (2012) 180.

401274 Hydrogen Production on Metal-Ceramic Micro-Structured Catalysts for Fuel Cell Applications

Abstract:

The heat and mass transfer properties of heterogeneous catalysts and their intrinsic kinetics are important factors that impact the performance of heterogeneous catalysts in practical applications. Such properties are crucial to the catalytic micro-reactors for hydrogen production for fuel cells, because the systems require intensive heat and mass flux for rapid process response under frequent daily start-up and shutdown (DSS) operation. In this work, we will present (i) the facile synthesis methods for core-shell metal-ceramic micro-composites with unique structural and catalytic properties and (ii) their catalyst applications in hydrogen production for fuel cells. Glycerol steam reforming to hydrogen on these catalysts was studied, which is an effective hydrogen production method for fuel cells utilizing a renewable biodiesel by-product.

15 AIChE
Spring Meeting, Austin, TX

Computational Thermodynamic Analysis of Solar Fuel Production via Metal Oxide Based H_2O and CO_2 Splitting Thermochemical Cycles

(Principal Author)
Dareen Dardor
Department of Chemical Engineering
College of Engineering
Qatar University, Doha, Qatar
dd1202223@student.qu.edu.qa

(Other Authors)
Rahul Bhosale
Department of Chemical Engineering
College of Engineering
Qatar University, Doha, Qatar
rahul.bhosale@qu.edu.qa

Anand Kumar
Department of Chemical Engineering
College of Engineering
Qatar University, Doha, Qatar
akumar@qu.edu.qa

Fares A. AlMomani
Department of Chemical Engineering
College of Engineering
Qatar University, Doha, Qatar
falmomani@qu.edu.qa

Shahd Gharbia
Department of Chemical Engineering
College of Engineering
Qatar University, Doha, Qatar
sg1000354@student.qu.edu.qa

Moustafa Hussein Ali
Department of Chemical Engineering
College of Engineering

Qatar University, Doha, Qatar
ma1203561@student.qu.edu.qa

Jamila Folady
Department of Chemical Engineering
College of Engineering
Qatar University, Doha, Qatar
jf1003686@student.qu.edu.qa

Shiva Yousefi
Department of Chemical Engineering
College of Engineering
Qatar University, Doha, Qatar
sy1106491@student.qu.edu.qa

Mehak Jilani
Department of Chemical Engineering
College of Engineering
Qatar University, Doha, Qatar
mj1104333@student.qu.edu.qa

Prepared for Presentation at
American Institute of Chemical Engineers
2015 Spring Meeting
11th Global Congress on Process Safety
Austin, Texas
April 27-30, 2015

UNPUBLISHED

Computational Thermodynamic Analysis of Solar Fuel Production via Metal Oxide Based H_2O and CO_2 Splitting Thermochemical Cycles

Dareen Dardor
Department of Chemical Engineering
College of Engineering
Qatar University, Doha, Qatar
dd1202223@student.qu.edu.qa

Rahul Bhosale, Anand Kumar, Fares A. AlMomani, Shahd Gharbia, Moustafa Hussein Ali, Jamila Folady, Shiva Yousefi, Mehak Jilani

Department of Chemical Engineering
College of Engineering
Qatar University, Doha, Qatar

Keywords: Solar Energy, Chemical Equilibrium Thermodynamics, Tin Oxide, Alternative Fuels, Thermochemical Redox Reactions, Second Law Analysis

Abstract

In this paper thermodynamic feasibility of the production of solar fuel via SnO_2/SnO based thermochemical CO_2 splitting redox reactions was investigated. The first step is the thermal dissociation of SnO_2 into SnO where concentrated solar energy is the source of this high temperature process heat. The second step which is a non-solar exothermic stage is the reaction of SnO with CO_2 to produce CO together with SnO_2 that is recycled back to the first step. Commercially available HSC Chemistry software and databases were used for the thermodynamic modeling of this process. Thermodynamic analysis was performed in two sections, namely, a) equilibrium thermodynamic calculations, and b) second law thermodynamic modeling of the SnO_2/SnO based thermochemical CO_2 splitting process. Results obtained via the computational thermodynamic modeling are presented.

1. Introduction

As the world population continues to grow with that comes a rapid increase in energy demands and an accompanying drastic depletion of fossil fuel reservoirs, it has thus become a critical issue to invest towards renewable energy technologies such as solar energy. A two-step solar thermochemical H_2O and/or CO_2 splitting process which utilizes metal oxide (MO) based redox reactions is one of the most promising ways for producing renewable fuels such as C, CO, H_2 or even a combination of CO and H_2 gas which is called syngas. Syngas is a precursor to solar fuels in the sense that by the Fischer Tropsch process, for example, it is possible to convert this syngas into liquid hydrocarbon fuels. Typically for high quality syngas the ratio of H_2:CO should be in the order of 2:1. The process of producing CO by such cycles for example includes the endothermic reduction of MOs at elevated temperature by releasing O_2 followed by the oxidation of MOs by CO_2 to produce CO.

Various MO based redox systems have previously been examined for the production of fuel via solar thermochemical H_2O and/or CO_2 splitting. Examples of these investigated cycles include ZnO/Zn [1-2], SnO_2/SnO [3-7], Fe_3O_4/FeO [8-9], mixed ferrites [10-20], and ceria based oxides [21-27]. In addition, several studies related to the thermodynamic analysis of MO based solar fuel production via thermochemical H_2O and CO_2 splitting reactions were also reported [1, 28-31]. Results obtained from previously reported findings indicate that among these MO systems, those based on the successive thermal dissociation and re-oxidation of volatile MOs e.g. SnO_2/SnO/Sn have been shown to exhibit the higher chemical reactivity.

Abanades et al. [3] suggested the two-step thermochemical water-splitting cycle based on SnO_2/SnO redox reactions. This study included an experimental investigation of the thermal reduction and water-splitting reactions in order to determine the optimum operating conditions, chemical conversion and H_2 yield [3]. The kinetics of the thermochemical conversion of H_2O into H_2 via SnO_2/SnO based redox reactions were examined by Chambon et al. [5]. Vishnevetsky and Epstein [6] studied the hydrolysis of different kinds of SnO_2 powders at different experimental conditions and moderate operating temperature range of 180 – 620°C. Recently, Abanades [7] investigated the reduction of CO_2 and H_2O by solar thermochemical looping using SnO_2/SnO redox reactions, the basis of the study was by performing thermogravimetric experiments.

In this study, thermodynamic analysis of the SnO_2/SnO based solar thermochemical CO_2 splitting process was investigated in two sections with the help of commercial thermodynamic HSC Chemistry software and databases. Firstly, chemical thermodynamic equilibrium of the SnO_2/SnO based solar thermochemical CO_2 splitting process was studied by examining the effect of various operating parameters on the thermal reduction and CO_2 splitting temperature. The second part of the study included the second law thermodynamic analysis of the process to determine its feasibility and efficiency. By performing the thermodynamic modeling of the process it was possible to determine; the solar absorption efficiency of the solar reactor ($\eta_{absorption}$), solar energy input to the solar reactor (Q_{solar}), net energy absorbed in the solar reactor ($Q_{reactor-net}$), radiation heat losses from the solar reactor ($Q_{re-radiation}$), rate of heat rejected to the surrounding from the quench unit (Q_{quench}), and from the SnO oxidizer ($Q_{sno-oxd}$), and solar to chemical energy conversion efficiency ($\eta_{solar-to-chemical}$) and results are presented in this paper. A typical SnO_2/SnO based two-step solar thermochemical CO_2 splitting process is presented in Figure 1.

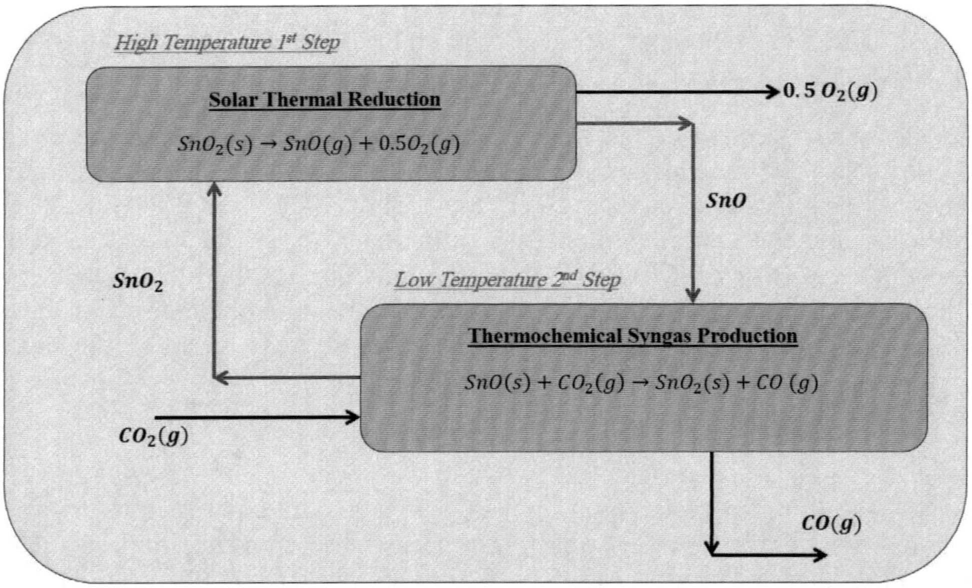

Figure 1. Tin oxide based solar thermochemical CO_2 splitting process.

2. Chemical Thermodynamic Equilibrium

2.1 Thermal reduction of SnO_2 into SnO

The thermal reduction of SnO_2 into SnO is represented in reaction (1).

$$SnO_2(s) \rightarrow SnO(g) + 1/2\, O_2(g) \tag{1}$$

This reaction can be summarized as the conversion of solid reactants into gaseous products. Effect of moles of inert gas (Ar) fed to the solar reactor on the thermal reduction temperature was examined in detail and obtained results are presented in Figure 2. The data reported shows that by increasing the molar flow rate of Ar in the feed it is possible to conduct the thermal reduction step at lower temperatures. For instance, at Ar = 10 moles/sec, the thermal reduction temperature was 2150°C. However, once the Ar moles were increased upto 230 moles/sec, the thermal reduction temperature reduced to 1783°C.

2.2 CO production via SnO_2/SnO based CO_2 Splitting

Reaction (2) represents the production of CO by oxidation of SnO via combined CO_2 splitting reaction.

$$SnO(s) + CO_2(g) \rightarrow SnO_2(s) + CO(g) \tag{2}$$

Figure 3 represents the change in the ΔG_{CO} with respect to the CO_2 splitting temperature. The results presented shows that at 300°C, ΔG_{CO} equals to zero which indicates that the production of CO by oxidation of SnO via CO_2 splitting reaction is feasible below 300°C.

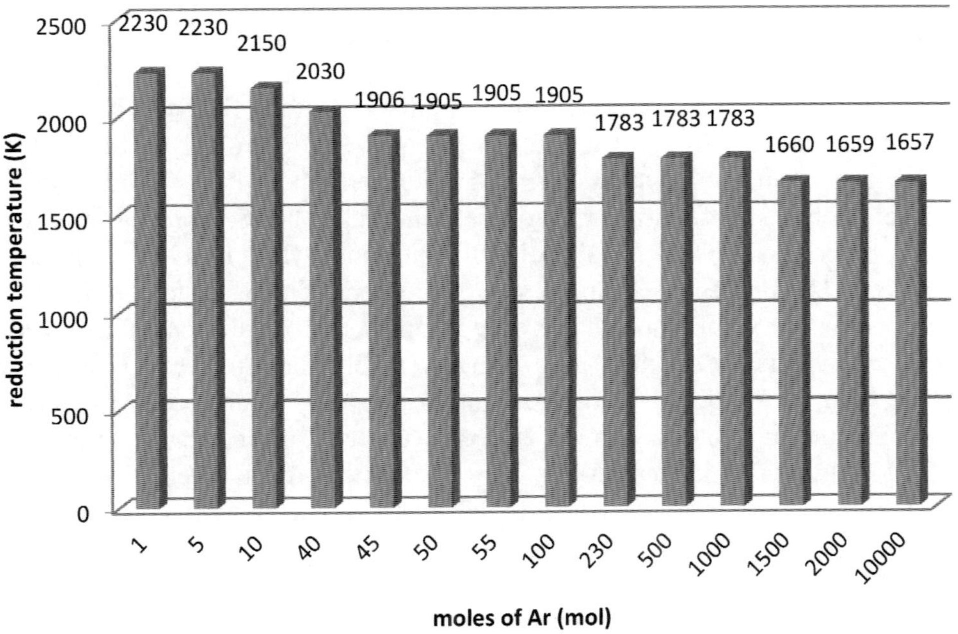

Figure 2. Effect of Ar feed (moles) on thermal reduction temperature of SnO_2(s) into SnO(g) and ½ O_2(g) [SnO_2 = 1 mole, P_{total} = 1 atm].

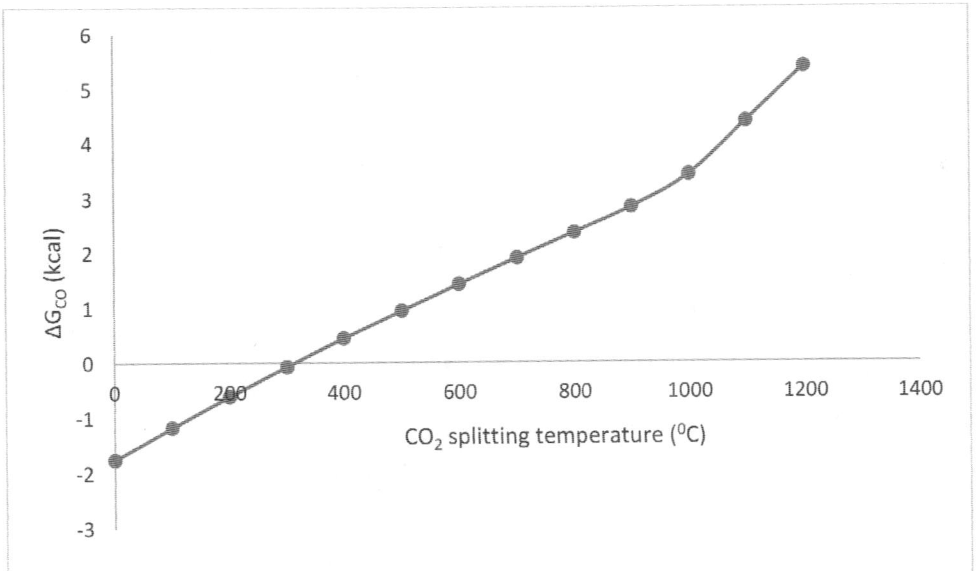

Figure 3. Effect of H_2O and CO_2 splitting temperature on ΔG_{CO}.

3 Second law thermodynamic analysis

3.1 *Process flow diagram for SnO_2/SnO based CO production process*

This section illustrates results of the second law analysis of the SnO_2/SnO based two-step solar thermochemical CO_2 splitting process. A process flow diagram for this solar fuel production operation is shown in Figure 4. The continuous SnO_2/SnO based CO production process consists of a solar reactor, a quench unit, a CO_2 splitting unit, a gas separator and a theoretical CO/O_2 ideal fuel cell. The molar flow rates of reactants are shown in Figure 4. The process is assumed to operate under steady state conditions at a total pressure of 1 atm. Some aspects have been omitted from the calculations these include heat exchangers used for recovering sensible latent heat. Furthermore, assumptions include: 1) perfectly insulated solar reactor 2) no convective/conductive heat losses, 3) kinetic and potential energies are neglected, 3) all reactions proceed to completion. Thermodynamic properties were extracted from HSC software and databases and the analysis is normalized to the SnO_2 molar flow rate (1 moles/sec) entering the solar reactor as shown in Figure 4.

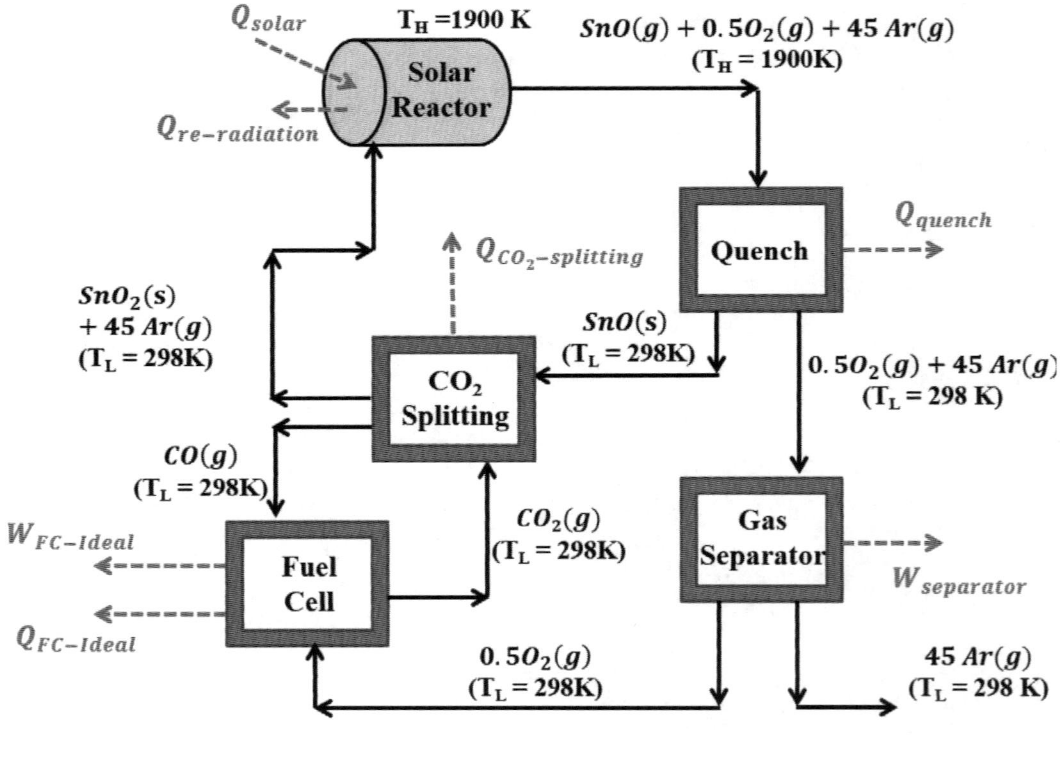

Figure 4. The process flow diagram of a two-step solar thermochemical SnO_2/SnO based CO production process.

3.2 Solar energy absorption efficiency ($\eta_{absorption}$)

$\eta_{absorption}$ is defined as the ratio of the net rate at which energy is being absorbed by the solar reactor divided by the solar power input through the aperture of the solar reactor and can be calculated using the following formula

$$\eta_{absorption} = \frac{Q_{reactor-net}}{Q_{solar}} = 1 - \left(\frac{\sigma T_H^4}{IC}\right) \tag{3}$$

Where, σ = Stefan-Boltzmann constant, I = normal beam insolation $1000(W/m^2)$, C= solar flux concentration ratio of the solar concentrating system (1000 suns), $Q_{reactor-net}$= net energy absorbed in the solar reactor, Q_{solar} = solar energy input. For this process at T_H = 1900K the $\eta_{absorption}$ = 26.1%.

3.3 Net solar energy absorbed by the solar reactor ($Q_{reactor-net}$)

According to the process flow diagram shown in Figure 4 the molar flowrate of entering SnO_2 is 1 mol/sec along with 45 mol Ar/sec both at 298K. By utilizing solar energy, these streams are heated up to 1900K and SnO_2 then undergoes 100% thermal reduction hence the net solar energy absorbed by the solar reactor can be calculated as:

$$Q_{reactor-net} = Q_{SnO_2-heating} + Q_{SnO_2-reduction} + Q_{Ar-heating} \tag{4}$$

The conducted thermodynamic calculations indicate that the energy required for the heating of SnO_2 is 143 kW in addition to the 1499 kW required for the heating of Ar. Finally, the amount of heat necessary for the reduction of SnO_2 is 547 kW and the total energy absorbed by the solar reactor amounts to 2198 kW.

3.4 Solar energy input to the solar reactor (Q_{solar})

For the obtained $\eta_{absorption}$ and $Q_{reactor-net}$, the total solar energy input needed to operate the solar reactor (Q_{solar}) for a CO production process by SnO_2/SnO based redox reactions can be determined by equation (5) and observed to be equal to 8384.86 kW.

$$Q_{solar} = \frac{Q_{reactor-net}}{\eta_{absorption}} \tag{5}$$

3.5 Radiation heat losses from the solar reactor ($Q_{re-radiation}$)

Radiation heat losses from the solar reactor conducting the thermal reduction of SnO_2 can be calculated by equation (6):

$$Q_{re-radiation} = Q_{Solar} - Q_{reactor-net} \tag{6}$$

For the solar reactor conducting the SnO_2/SnO based solar thermochemical CO_2 splitting process, the $Q_{re-radiation}$ equals to 6196.41 kW.

3.6 *Rate of heat rejected to the surrounding from the quench unit* (Q_{quench})

As shown in the process flow diagram illustrated in Figure 4, the gaseous products of the thermal reduction step are sent to a quench unit where they are rapidly cooled down from the thermal reduction temperature (T_H= 1900K) to ambient temperature (T_L= 298K). This quench unit is necessary to avoid recombination of the high temperature gaseous products. The rate of heat rejected to the surroundings by the quench unit (Q_{quench}) is calculated by Eq. 7 and observed to be equal 1891.53 kW.

$$Q_{quench} = -\dot{n}\Delta H|_{SnO(g)+0.5O_2(g)+45\mathrm{Ar}\,(g)@1900K \rightarrow SnO(s)+0.5O_2(g)+45\mathrm{Ar}\,(g)@298K} \quad (7)$$

3.7 *Rate of heat rejected to the surrounding from the CO_2 splitter* $(Q_{CO2-splitting})$

The solidified SnO leaving the quench is sent to the CO_2 splitting unit where it is reacted with a stream of CO_2 to form CO(g) and SnO_2(s). For this step, complete conversion of SnO to SnO_2 is assumed and the rate of heat rejected to the surrounding from the CO_2 splitter $(Q_{CO2-splitting})$ is estimated by Eq. (8) to be 13.96 kW.

$$Q_{CO2-splitting} = -\dot{n}\Delta H|_{SnO(s)+CO_2(g)\,@298\mathrm{K} \rightarrow SnO_2(s)+CO(g)@298K} \quad (8)$$

3.8 *Rates of theoretical work performed* $(W_{FC-Ideal})$ *and heat rejected* $(Q_{FC-Ideal})$ *by the fuel cell*

An ideal fuel cell is added to the process flow diagram in Figure 4 in order to complete the SnO_2/SnO based CO production cycle and calculate the maximum possible theoretical work that can be extracted from this cycle. The ideal fuel cell operates in the presence of CO/O_2, and the rates of theoretical work performed $(W_{FC-Ideal})$ and the heat rejected by the fuel cell $(Q_{FC-Ideal})$ can be calculated by using equations (9) and (10) where the efficiency of the fuel cell (η_{FC}) is assumed to be 100%.

$$W_{FC-Ideal} = -(\eta_{FC}) \times \dot{n}\,\Delta G|_{CO(g)+0.5O_2(g)\rightarrow CO_2(g)@298K} \quad (9)$$

$$Q_{FC-Ideal} = -(T_R) \times \dot{n}\,\Delta S|_{CO(g)+0.5O_2(g)\rightarrow CO_2(g)@298K} \quad (10)$$

For the SnO_2/SnO based solar thermochemical CO production process described in Figure 4, the $W_{FC-Ideal}$ and $Q_{FC-Ideal}$ were observed to be equal to 257.35 kW and 25.766 kW, respectively.

3.9 *Solar to chemical energy conversion efficiency* $(\eta_{solar-to-chemical})$

This quantity is defined as the portion of solar energy that is converted into chemical energy (i.e. by the Gibbs free energy of the products) or in other words the maximum possible work that can be extracted from the products.

$$\eta_{solar-to-chemical} = \frac{W_{FC-Ideal}}{Q_{solar}} \quad (11)$$

The $\eta_{solar-to-chemical}$ of the SnO_2/SnO solar thermochemical CO production process is calculated using equation (11) and is equal to 3.37%.

4. Conclusions

This study involved the thermodynamic examination of two-step solar thermochemical cycles for CO production via SnO_2/SnO based CO_2 splitting reactions. Thermodynamic equilibrium calculations performed at a pressure of 1 atm indicated that by increasing the concentration of inert Ar gas in the feed stream the required temperature for the thermal reduction of SnO_2 into SnO can be significantly decreased. According to the performed calculations, it was further observed that the CO_2 splitting via SnO oxidation is feasible below 300°C. The second law thermodynamic analysis of the solar reactor conducting the thermochemical SnO_2/SnO based CO_2 splitting process at 1900 K and at a solar concentration of 1000 suns indicate $\eta_{absorption}$ = 26.1%, Q_{solar}=8384.86 kW, $Q_{reactor-net}$ =2198 kW, $Q_{re-radiation}$ =6196.41 kW, Q_{quench}=1891.53 kW, $Q_{CO2-splitting}$ =13.96 kW, and $\eta_{solar-to-chemical}$ = 3.37%.

5. Acknowledgments

The authors gratefully acknowledge the financial support provided by the Qatar University Internal Grants QUUG-CENG-CHE-13/14-4 and QUUG-CENG-CHE-14\15-10.

References

[1] Loutzenhiser, P., Steinfeld, A.," Solar syngas production from CO_2 and H_2O in a two-step thermochemical cycle via Zn/ZnO redox reactions: Thermodynamic cycle analysis", *Int. J. Hydrogen Energy*, Volume 36, page No. 12141, 2011.

[2] Stamatiou, A., Loutzenhiser, P., Steinfeld, A., "Solar Syngas Production from H_2O and CO_2 via Two-Step Thermochemical Cycles Based on Zn/ZnO and FeO/Fe_3O_4 Redox Reactions: Kinetic Analysis", *Energy Fuels,* Volume 24, page No. 2716, 2010.

[3] Charvin, P., Abanades, S., Lemont, F., Flamant, G., "Experimental study of SnO2/SnO/Sn thermochemical systems for solar production of hydrogen", *AIChE J.*, Volume 54, page No. 2759, 2008.

[4] Abanades, S.," CO_2 and H_2O reduction by solar thermochemical looping using SnO2/SnO redox reactions: Thermogravimetric analysis", *Int. J. Hydrogen Energy,* Volume 37, page No. 8223, 2012.

[5] Chambon, M., Abanades, S., Flamant, G.," Kinetic investigation of hydrogen generation from hydrolysis of SnO and Zn solar nanopowders", *Int. J. Hydrogen Energy,* Volume 34, page No. 5326, 2009.

[6] Vishnevetsky, I., Epstein, M., J. *Sol. Energy Eng.* 131 (2009) 021007-1.

[7] Abanades, S., Villafan-Vidales, H.," CO_2 and H_2O conversion to solar fuels via two-step solar thermochemical looping using iron oxide redox pair", *Chem. Eng. J.*, Volume 175, page No. 368, 2011.

[8] Gokon, N., Murayama, H., Nagasaki, A., Kodama, T.," Thermochemical two-step water splitting cycles by monoclinic ZrO_2-supported $NiFe_2O_4$ and Fe_3O_4 powders and ceramic foam devices", *Sol. Energy,* Volume 83, page No. 527, 2009.

[9] Scheffe, J., Li, J., Weimer, A.," A spinel ferrite/hercynite water-splitting redox cycle", *Int. J. Hydrogen Energy,* Volume 35, page No. 3333, 2010.

[10] Roeb, M., Gathmann, N., Neises, M., Sattler, C., Pitz-Paal, R.," Thermodynamic analysis of two-step solar water splitting with mixed iron oxides", *Int. J. Hydrogen Energy,* Volume 33, page No.893, 2009.

[11] Lorentzou, S., Agrafiotis, A., Konstandopoulos, A.," Aerosol spray pyrolysis synthesis of water-splitting ferrites for solar hydrogen production", *Granular Matter*, Volume 10, page No. 113, 2008.

[12] Bhosale, R., Shende, R., Puszynski, J., "H_2 generation from thermochemical water-splitting using sol-gel derived Ni-ferrite", *Journal of Energy & Power Engineering*, Volume 4, Page No. 27, 2010.

[13] Bhosale, R., Shende, R., Puszynski, J., "H_2 generation from thermochemical water-splitting using sol-gel synthesized Zn/Sn/Mn-doped Ni-ferrite," *International Review of Chemical Engineering*, Volume 2, Page No. 852, 2010.

[14] Bhosale, R., Khadka, R., Shende, R., Puszynski, J., "H_2 generation from two-step thermochemical water-splitting reaction using sol-gel derived SnxFeyOz", *Journal of Renewable & Sustainable Energy*, Volume 3, Page No. 063104-1, 2011.

[15] Bhosale, R., Shende, R., Puszynski, J., "Thermochemical water-splitting for H_2 generation using sol-gel derived Mn-ferrite in a packed bed reactor", *International Journal of Hydrogen Energy*, Volume 37, Page No. 2924, 2012.

[16] Bhosale, R., Shende, R., Puszynski, J.," Sol-Gel Derived $NiFe_2O_4$ Modified with ZrO_2 for Hydrogen Generation from Solar Thermochemical Water-Splitting Reaction", *Mater. Res. Soc. Symp. Proc,* Volume 1387, 2012.

[17] Bhosale, R., Alxneit, I, van den Broeke, Leo L. P., Kumar, A., Jilani, M., Gharbia, S., Folady, J., Dardor, D.," Sol-Gel Synthesis of Nanocrystalline Ni-Ferrite and Co-Ferrite Redox Materials for Thermochemical Production of Solar Fuels", *Mater. Res. Soc. Symp. Proc.*, Volume 1657, 2014.

[18] Bhosale, R., Shende, R., Opoku, M. K. Puszynski, J., *AIChE Annual Meeting Proc.*, 2009.

[19] Bhosale, R., Shende, R., Puszynski, J., *AIChE Annual Meeting Proc.*, 2010.

[20] Pasala, X., Bhosale, R., Amar, V. S., Puszynski, J., Shende, R., *AIChE Annual Meeting Proc.*, 2012.

[21] Shende, R., Puszynski, J., Opoku, M., Bhosale, R.," Synthesis of Novel Ferrite Foam Material for Water-Splitting Application", *NSTI-Nanotech Conference & Expo. Proc.*, Volume 1, page No. 201, 2009.

[22] Abanades, S., Flamant, G.," Thermochemical hydrogen production from a two-step solar-driven water-splitting cycle based on cerium oxides", *Sol. Energy,* Volume 80, page No. 1611, 2006.

[23] Chueh, W. C., Falter, C., Abbott, M., Scipio, D., Furler, P., Haile, S., Steinfeld, A.," High-Flux Solar-Driven Thermochemical Dissociation of CO_2 and H_2O Using Nonstoichiometric Ceria", *Science,* Volume 330, page No. 1797, 2010.

[24] Kaneko, H., Miura, T., Ishihara, T., Taku, S., Yokoyama, T., Nakajima, H., Tamaura, Y.," Cerium ion redox system in CeO_2–xFe_2O_3 solid solution at high temperatures (1,273–1,673 K) in the two-step water-splitting reaction for solar H_2 generation", Energy, Volume 43, page No. 3153, 2008.

[25] Meng, Q., Lee, C., Ishihara, T., Kaneko, H., Tamaura, Y.," Reactivity of CeO_2-based ceramics for solar hydrogen production via a two-step water-splitting cycle with

concentrated solar energy", *Int. J. Hydrogen Energy,* Volume 36, page No. 13435, 2011.

[26] Le Gal, A., Abanades, S.," Catalytic investigation of ceria-zirconia solid solutions for solar hydrogen production", *Int. J. Hydrogen Energy*, Volume 36, page No. 4739, 2011.

[27] Abanades, S., Gal, A. L.," CO_2 splitting by thermo-chemical looping based on $Zr_xCe_{1-x}O_2$ oxygen carriers for synthetic fuel generation", *Fuel,* Volume 102, page No.180, 2012.

[28] Le Gal, A., Abanades, S., Flamant, G.," CO_2 and H_2O Splitting for Thermochemical Production of Solar Fuels Using Nonstoichiometric Ceria and Ceria/Zirconia Solid Solutions", *Energy Fuels,* Volume 25, page No. 4836, 2011.

[29] Scheffe, J., Steinfeld, A.," Solar Thermochemical CO_2 Splitting Utilizing a Reticulated Porous Ceria Redox System", *Energy Fuels,* Volume 26, page No. 1928, 2012.

[30] Galvez, M. E., Loutzenhiser, P. G., Hischier, I., Steinfeld, A.," CO_2 Splitting via Two-Step Solar Thermochemical Cycles with Zn/ZnO and FeO/Fe_3O_4 Redox Reactions: Thermodynamic Analysis", *Energy Fuels,* Volume 22, page No. 3544, 2008.

[31] Galvez, M. E., Frei, A., Albisetti, G., Lunardi, G., Steinfeld, A.," Solar hydrogen production via a two-step thermochemical process based on MgO/Mg redox reactions—Thermodynamic and kinetic analyses", *Int. J. Hydrogen Energy,* Volume 33, page No. 2880, 2008.

401275 Catalytic Glycerol Steam Reforming to Hydrogen on Metal-Ceramic Composites

Abstract:

Glycerol steam reforming on Ru and Rh catalysts supported on metal-ceramic composites is studied at high temperature and space velocity conditions for fuel cell applications. The catalysts exhibit substantially enhanced fuel conversion to hydrogen compared to the conventional catalyst, which indicates its superior properties for the reaction due to the fine active metal dispersion and facilitated heat and mass transport contributed from the unique catalyst microstructures.

398091 Synthesis and Applications of Nanocoatings

Abstract:

The production of cell phones and emergence of solar power plants has lead to the preparation of nanocoatings. Flake production, functionalization, particulation, sacrificial coating, specifications are critical in order to prepare lubricous, high performance, nanocoatings. Questek developed a steel grade made of nanophase precipitated dispersion. The three phases during the development of nanocoating are; (i) conception, production in bench scale, scale-up to pilot an plant; (ii) obtaining license, purpose of development of product and; (iii) marketing. Coating is the interface between product and environment.

GMR, giant magneto resistive thin film heads used in hard drives of desktop computers comprises of seven layers. The seven layers are silicon substrate, tantalum buffer layer, nickel-iron free layer, copper spacer, germanium pinned layer, iron-manganese pinning layer and tantalum gap layer. In addition to GMR, nanocoatings are also used as thermal barrier coatings. Magnesium oxychloride sorrel cement coatings are used, for example. Ceramic coatings can be used to prevent corrosion and oxidation of blades. Electron beam physical deposition process is used to obtain nanocoating. SSA, specific surface area can be used to characterize nanocoating such as the degree of nanostructuring. TBC comprises of oxide layer, bond coat layer, insulating layer, mixed oxide particles with particle size in the range of 10-100 nm. Aluminum projections with width of 1-5 nm and lateral grains with radius of curvature less than 1 A and length 5-8 nm are other nanofeatures. TBC can be used to prevent fire spread. Transient temperature in TBC was modelled including the effects of damped wave conduction and relaxation. There is a critical thickness of coating below which subcritical damped oscillations in temperature can be expected. For materials with relaxation times greater than $R^2/15.33a_{nr}$ the temperature in the naval warhead can be expected to undergo subcritical damped oscillations.

X-ray lithography and electroplating techniques were used in order to create a zone plate of thickness of 500 nm with zone dimensions less than 50 nm. PU coating can be grafted with nano-alumina in order to improve the performance of the coating by increasing the adhesion between nanoalumina and organic substrate. Coating machines and rollers can be used to manufacture solar cells at lower cost. Flexible film are laminated on top of cell phones. A PV, photovoltaic device was created by casting a solution of conducting polymer and fullerenes into a thin film. Nanocoatings has been shown to improve the light to electricity conversion efficiency in solar cells.

Nanostructured alumina/titania powders can be used to improve abrasive wear resistance of ceramic coatings by 400%. Process comprises of dispersion of nanopowders into a colloid, binded and granules spray dried using a Metro 9 MB plasma gun. Thermal spray robots are used to obtain nanocoating. Nanostructures can result from control of morphology of phase separated polymer blends. Kinetics and thermodynamics of phase separation can be modeled. Polymer morphology with a single characteristic length scale can be made by quenching partially miscible polymer blend below the critical temperature of demixing. Spinodal patterns are formed that evolve with passage of time. Nanostructuring can be controlled during preparation of metal chalcogenide nanostructured films by electrodeposition. These films can be formed within mesopores of silica.

Cost of solar cells cen be reduced by use of graphene layers as AR coatings. Light to electricity conversion was found to increase to 14.6%. Graphene sheets with less than 24,000 atoms/25 nm are metastable. This can result when Gibbs free energy of mixing is negative and the phase stability criterion is not met. Large single layer graphene sheets that are made by roll transfer are stable. Annular Plug Flow Reactor is used to decompose acetylene autocatalytic decomposition and carbon is grown by deposition on rolls.

Nanocoatings are used in the design and application of nanorobots for drug delivery. Principles of photodynamic therapy, fullerene chemistry, nanostructuring, x-rays, computers, pharmacokinetic and robotics are applied to design a nanorobot in order to treat Alzheimer's disease. Drug is inactive in caged form and activiated by breakage of fullerenes by irradiation at desired site.

Protein folding, secondary structure of proteins, nucleic acid structure, ribosomes, chromosomes and phospholipid membranes are examples where self-assembly is critical.

Graphene is a interesting material of choice in Naval applications in order to demonstrate an infrared-transparent, electromagnetic shielding coating that can be applied to electrooptic sensor windows and domes. Transmission greater than 90% in the 3-5 micron wavelength region can be achieved

Fluidized bed can be used to obtain the nanocoating. The precursor material is calcined in ammonia at 1000 ^{0}C for 48 hours in a fluidized bed reactor. A red tri-tantallumpenta nitride, Ta_3N_5 coated mica pigment is produced. Bubbles and slugs and other instabilities formed during the operation of the fluidized bed are reduced by use of vibrators/agitators. The substrate particle size range is between 1 – 500 mm.After the pigments of the desired thinness is obtained a surface treatment step can be used to increase the weather stability and chemical stability of the pigment

OVERCOMING KINETIC LIMITATIONS OF Cr(VI) ASORPTION ONTO BIOSORBENTS: BIOMAS-MAGNETITE BIONANOCOMPOSITE

Pholosi A., Ofomaja A.E., Naidoo E.B.

Biosorption and Water Treatment Research Laboratory, Department of Chemistry,

Faculty of Applied and Computer Sciences, Vaal University of Technology, P. Bag X021,

Vanderbijlpark, 1900 South Africa.

INTRODUCTION

Biosorbents are a promising class of adsorbents discovered in the last decade for their potential application in water and waste water treatment **(Abdolali et al., 2014)**. One class of biosorbents are the lignocellulosic materials obtained from plant wastes or by-products (**Ngah and Hanafiah, 2008).** Lignocellulosic materials are composed of macromolecules such as lignin, cellulose and hemicellulose compounds which are known to contain groups that exhibits exchange and complexating properties **(Ofomaja and Ho, 2007)**. The lignin portion of lignocellulosics acts a binding agent holding the celluloses fraction to the hemicellulose fraction and keeps the plant materials erect and firm. As a result particles of biosorbents materials applied as adsorbents possess very low porosity and internal surface causing restriction of fluid flow.

This feature of biomasses, as they are called generally leads to several challenges when during application of biomasses as adsorbents and these includes (i) low diffusion rates of pollutants within the adsorbent particles during adsorption **(Thinh et al., 2013)**, (ii) high operating pressures when applied in packed columns **(Yavuz et al., 2006)**, low adsorption rates and capacities **(Thinh et al., 2013)**. To overcome these shortcomings researchers have resorted to reducing the sizes of the biomass materials by crushing or grinding them to smaller particles sizes. Small particle sized materials for adsorptions are also known to come with its challenge, which is the separation of the adsorbent particles from the treated solution at the end of contact **(Safarik et al., 2007)**.

Currently researchers have turned to the use of bio-nanocomposites consisting biomaterials coated with magnetic nanoparticles to overcome these limitations. Although much success have been achieved in applying these novel biomaterials composites yet little is known about how and to what extent these limitations are dealt with. This paper therefore seeks to study how and to what extent to which kinetic limitations are avoided when pine biomass coated with magnetite is applied as adsorbent for Cr(VI) from aqueous solution.

METHODS:

Pine cone was treated with 0.15 mol/dm^3 NaOH and the 1.0 g of the product added to a solution containing a mixture of Fe^{2+}/Fe^{3+} of molar ratio 2:1. Magnetic was then precipitated onto the pine surface using NH_4OH. The biocomposite and the raw pine biomass were characterized. Batch kinetic adsorption studies of Cr(VI) from aqueous solution was performed using both materials from a 150 mg/dm^3 solution at 26, 31, 36, 41 and 46 ^{o}C. Three nonlinear kinetic and three diffusion models were applied to test the kinetic data.

RESULTS:

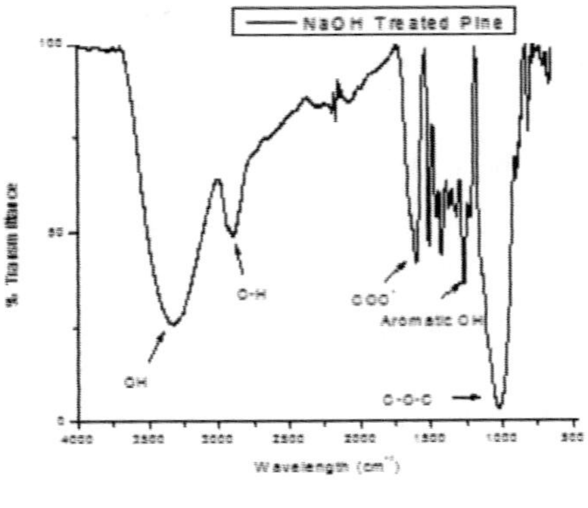

FTIR spectra for the NaOH treated and MNP-Pine composite are shown in Figs. 1 and 2. The presence of Fe-O and Fe-OH bonds in the composite confirms that presence of magnetite in the bio-nanocomposite.

Fig. 1: FTIR of NaOH treated pine

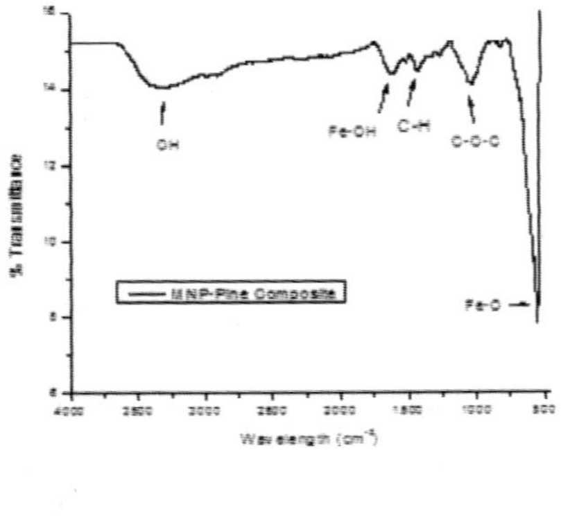

Fig. 2: FTIR of MNP-Pine Composite.

Surface characteristics of the NaOH treated and the bio-composite are shown in Table 1. The results shows that an increase of BET surface from 2.67 m^2/g to 68.91 m^2/g in the bio-composite. The external surface and pore width were also increased drastically with the composite. The small size of the magnetite coating as well as the possibility of magnetite formation within the pores of the pine causing expansion and breakup of the pine particles may be responsible for this happening.

The kinetic data for the adsorption of Cr(VI) onto NaOH and MNP-Pine composite were fitted to the intraparticle diffusion model in Fig. 3. The results indicate the while the metal uptake per square-root of time profile for the NaOH treated pine can be broken into two sections, the MNP-Pine composite profile can be broken into three sections.

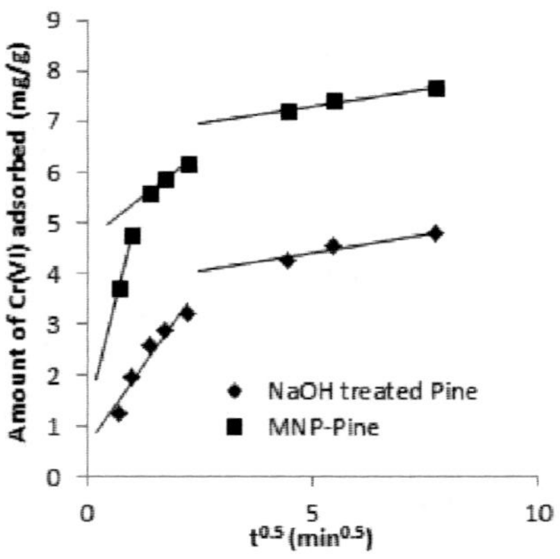

Fig. 3: Intraparticle Diffusion plots for NaOH treated pine and MNP-Pine.

The implication of this is the decrease in kinetic restriction by the creation of increased internal surface in the composite. The intraparticle diffusion constants were found to be higher for the bio-composite than for the NaOH treated pine signifying free flow of aqueous solution through the internal surface of the material. The relationship between intraparticle diffusion constant and temperature is shown in Table 2. Also it was observed that the constants increased with increasing solution temperature. Increasing solution temperature increases the rate of migration which is responsible for the increased diffusion of the pollutant through the adsorbent material.

CONCLUSION:

Kinetic limitations associated with the use of biomaterials as adsorbents can be eliminated by coating of biomaterials with nanoparticles. The resulting bio-composite processes better surface properties and improved diffusion properties for pollution removal from aqueous solution.

References

Safarik, I., Lunackova, P., Mosiniewicz-Szablewska, E., Weyda, F., Safarikova, M. Adsorption of water-soluble organic dyes on ferrofluid-modified sawdust. Holzforschung 61 (2007) 247-253.

Yavuz, H., Denizli, A., Güngünes, H., Safarikova, M., Safarik, I. Biosorption of mercury on magnetically modified yeast cells. Separation and Purification technology 52 (2006) 253-260

Abdolali, a., Guo, W.S., Ngo, H.H., Chen, S.S., Nguyen, N.C., Tung, K.L. Typical linocellulosic wastes and by-products for biosorption process in water and wastewater: Bioresource Technology 160 (2014) 57-66.

Ngah.,W.S.W., Hanafiah, M.A.K.M. Removal of heavy metal ions from wastewater by chemically modified plant waste as adsorbents: A review. Bioresource Technology 99 (2008) 3935-3948.

Table 1: Surface Properties

Properties	NaOH treated Pine	MNP-Pine Composite
Surface area (m^2/g)	2.67	68.91
Pore volume (cm^3/g)	0.002	0.153

External surface (m^2/g)	2.25	48.79
Ave. pore width (nm)	8.89	33.10

Table 2: Diffusion parameters for Cr(VI) adsorption onto NaOH treated and MNP-Pine at different temperatures

Temp. (^oC)	Intraparticle diffusion rate constant (mg/g $min^{0.5}$)		
	NaOH Treated Pine	MNP-Pine Composite	
	k_{i1}	k_{i1}	k_{i2}
26	0.1465	0.1404	0.6828
31	0.1994	0.2192	0.8464
36	0.2450	0.2662	1.0963
41	0.2690	0.3142	1.3046
46	0.3110	0.3789	1.7028

399757 Ni-Cu Based Nano Composite Alloys for Methanol Electro-Oxidation Reaction

Abstract:

PEMFC production coast is one of the important issues for its commercialization. Platinum (electro catalyst) and Nafion (membrane), both are very expensive. Efforts have been made in recent years to reduce catalyst costs and improve the efficiency of non-platinum based catalysts for both cathode and anode application. In the case of direct methanol fuel cell (DMFC), most efficient anode catalyst for the electrochemical oxidation reaction is the platinum nano-particles (2 - 4 nm) supported on a high surface area carbon carbon nanotubes (CNTs). However, it has been found that carbon monoxide (CO) formed during the reaction as a by-product, poisons the surface of the anode which in turn causes a dramatic decrease in the anode performance. A possible solution for that will be to use a CO tolerant electro catalyst and it is tried by alloying Pt with other transition metals or using non-platinum based catalysts. It is reported that the best CO-tolerant electro catalysts is Pt-Ru alloy. Again, this alloy is also extremely expensive, so there is a need for an alternative low coat electro catalyst with desired performance. Many non-platinum based cheaper catalysts have been investigated. However, due to the inferior performance of these catalysts, there is a big opening to develop low-cost non-platinum multi composition electro catalysts with desired properties for DMFC/PEMFC. Here, we will report synthesis; characterization and methanol electro-oxidation characteristics of several novel Ni-Cu based multi-composition alloy catalysts on MWCNT support.

401509 The Enhancement of Liquid-Liquid Extraction Performance By Using Hollow Oil Droplets with Microbubbles As Dispersed Phase (Electronic Poster Presentation)

Abstract:

The Enhancement of Liquid-liquid Extraction Performance by Using Hollow Oil Droplets with microbubbles as Dispersed Phase

Wenting Wang, Guotao Liu, Jing Tan, Jianhong Xu*, Guangsheng Luo

State Key Laboratory of Chemical Engineering, Department of Chemical Engineering, Tsinghua University, Beijing 100084, China

Micro-structured chemical systems have become one of most important methods for realizing greener, more efficient, safer and more controllable chemical engineering processes, and they continue to show promise for the further development of modern chemical industries. In this work, process intensification of liquid¨Cliquid extraction with high phase ratio by forming G/W/O (gas in water phase in organic phase) microdispersion system is carried out.

Extraction of solute from an organic phase to an aqueous phase was selected as the model system. The influence of the addition of gas phase as well as dispersion size on mass transfer performance is investigated. The results show that both the increase in the addition of gas phase and decrease in dispersion size lead to significant enhancement in mass transfer performance, especially for the overall mass transfer coefficient. As gas-to-water flow rate ratio increases from 25 to 600, the overall volumetric mass transfer coefficient increases by nearly two orders of magnitude. The calculation result shows good coincidence with experimental values. The modeling investigation possibly leads to a further understanding of the mechanism of process intensification of liquid¨Cliquid extraction process by introducing microbubbles and forming G/W/O microdispersion system and also helps the development of new extraction processes as well as design of microdispersion devices with compactness and high efficiency.

Key words: micro-dispersion; G/O/W double emulsion; liquid- liquid extraction; mass transfer

396369 Experimental Investigation of Oil Recovery By Hot Water Followed By Steam Flooding for Lower Fares Heavy Oil

Abstract:

In this study, a laboratory hot water and steam flooding tests were conducted. The tests incorporate five core samples from Lower Fares, unconsolidated sandstone. The hot water flood tests were conducted at 60 and 80 degrees Celsius. Both steam displacement and vertical steam displacement tests were conducted at 153 degrees Celsius. Several factors effecting oil recovery such as; injection pressure, injection rate, and injection temperature were investigated.

The main objective of this work is to study the recovery of heavy oil for Lower Fares reservoir by hot water injections at various temperatures followed by steam injection. Numbers of runs wereconducted in horizontal and vertical cores to obtain analytical data. Results of this experiment is used to determine the optimum recovery method for heavy oil. Both core samples and fluid samples were tested at surface conditions. The results indicate an average cumulative oil recovery of 26.5% at 60°C, 44.7% at 80°C, and ultimately increase to 68% by horizontal steam displacement and 77.5% by vertical steam displacement (VSD) method.

401098 Mitigation of the Cytotoxicity of Single-Walled Carbon Nanotubes Utilizing Surface Functionalization and Nontoxic Surfactants

Abstract:

This research sought to address carbon nanotubes' (CNT) toxicity when released into the environment through the addition of non-toxic surfactants and the novel method of surface functionalization. It is important that the cytotoxicity of CNTs be understood because as their usage increases, the likelihood of their release increases. This can have direct negative effects on the health of both terrestrial organisms – including humans – and aquatic organisms. In testing how the two different methods of solubizing single-walled carbon nanotubes (SWNT) affects cytotoxicity, it was predicted that both methods would decrease cytotoxicity, but that the surfactant suspension method would be more effective. The effects at different concentrations of each nanotube treatment were tested on the growth rates of *P. subcapitata* and the oxidative stress on *E. coli* K12. While both methods deceased cytotoxicity, the novel method of surface functionalization significantly outperformed the addition of surfactants, shown by an average increase of 55% in the growth rate of *P. subcapitata* and an average decrease of 20% in the oxidative stress of *E. coli* when surface functionalization was used instead of surfactants. Based on these findings, it is suggested that surface functionalization of CNTs should be implemented when risk of environmental contamination is apparent.

399276 Enantioseparation of Phenylethanol By Stripping Crystallization

Abstract:

Stripping crystallization (SC) is a new separation technique which combines distillation and crystallization. Basically, SC is operated at a triple-point condition, in which the liquid is simultaneously vaporized and crystallized due to the three-phase equilibrium. In essence, the process is continued until the liquid phase is completely eliminated and only the pure solid crystals remain in the feed. Thus, crystal washing is not required since only pure solid crystals remain in the feed and no impurities are adhered on the crystal surfaces at the end of SC. In the previous research, SC has been successfully applied to separate p-xylene from the mixed xylenes by Shiau et al. (2005, 2008).

Phenylethanol is an important chiral compound with a number of potential applications. These alcohols are used as building blocks for the synthesis of bioactive compounds such as pharmaceuticals, agrochemicals, and natural products. S-(+)- phenylethanol can be used to treatment of E. coli and other synthetic drugs precursors. A thermodynamic model is developed to simulate the three-phase equilibrium conditions during the SC operation. The experiments show that seeding with magnetic stirring can be efficiently employed to purify S-phenylethanol from phenylethanol enantiomers. The experimental results, including the final enantiomeric purity and recovery ratio of S-phenylethanol, are consistent with the simulation results predicted by the model.

1. Shiau, L. D., Wen, C. C. and Lin, B. S. Separation and purification of p-xylene from the mixture of m-xylene and p-xylene by distillative freezing, *Ind. Eng. Chem. Res*, **44(7)**, 2258- 2265, 2005.

2. Shiau, L. D., Wen, C. C. and Lin, B. S. Separation of p-xylene from the multicomponent xylene system by stripping crystallization, *AIChE Journal*, **54(1)**, 337-342, 2008.

399678 A Quantum Theoretical Development of Platinum-Ruthenium-Tin Anodic Catalysts for Enhanced Direct Ethanol Fuel Cell Performance

Abstract:

In pursuit of clean and renewable energy, fuel cell technology studies aim to combat growing global energy concerns. The design of anodic catalysts for direct ethanol proton exchange membrane fuel cells has been a particular focus to offer goals of enriched efficiency, longevity, and cost. Computational efforts in surface chemistry allow for a semi-quantitative description of reaction processes to be developed that will aid in meeting said objectives. Presently, the assessment of metal alloying effects on the activity of Pt-based nanostructures for the electro-oxidation of ethanol appears promising. This study employs the DMol^3 suite of the Materials Studio 6.0 software to conduct First-Principles, Density Functional Theory and Transition State Theory, calculations as a theory-driven approach to outlining equilibrium Pt-Ru-Sn (2:2:1) nanostructures and exploring the ethanol oxidation reaction (EOR) mechanism across the catalyst surface. A systematic optimization is performed on two candidate configurations with functionals of Local-Density Approximation and Generalized-Gradient Approximation exchanges accompanied by varied correlation combinations. Lastly, an exploration of EOR transition states is carried out using a complete Linear Synchronous Transit/Quadratic Synchronous Transit method, prefaced by a similar optimization procedure on species-catalyst complexes. Results shown include energy profiles, itemized by: (1) Cohesive Energy, E_c, of 10-atom structure and (2) Activation Energy, E_a, for EOR across metal surface. This information provides a framework for understanding important kinetics and thermodynamic details that could lead to the development of novel catalysts for energy applications.

399760 Enhanced Modification Effect of CeO2 on Pt-Pd Binary Catalysts for Formic Acid Oxidation

Abstract:

This article deals with the promotional effects of CeO_2 on PtPd/CeO_2-OMC electrocatalysts. The synthesized catalysts are characterized using different physicochemical techniques and evaluated in a formic acid oxidation fuel cell. N_2 adsorption/desorption analysis shows that CeO_2 modification increases the surface area of OMC from 1005 m^2/g to 1119 m^2/g. SEM, XRD and TEM analysis reveal that the presence of CeO_2 enhances the active metal(s) dispersion on the CeO_2-OMC surface. The average particle size of the dispersed metal decreases with the increase of Pt/Pd ratio on CeO_2-OMC support. Cyclic voltametry measurement of Pd/CeO_2-OMC gives 12 % higher anodic current activity with 83 mV negative shift of the peak potential as compared to unmodified Pd/OMC. In bimetallic catalysts, the addition of Pt improves the activity and stability of the catalysts significantly. Among the bimetallic samples, Pd_3Pt_1/CeO_2-OMC displays superior current density (74.6 mA/cm^2), which is 28.3 times higher than that of Pt/CeO_2-OMC. It also shows higher stability in extended period of runs with least indication of CO poisoning effects.

399949 Topology Optimisation for Optimal Shape Selection of Solar Parabolic Trough Receiver (Electronic Poster Presentation(Electronic Poster Presentation (Electronic Poster Presentation)

Abstract:

Concentrated Solar thermal technology holds a strong potential for the generation of process heat in the chemical process and allied industries including residential and commercial application. For more than a century the shape of the receiver of a solar thermal parabolic trough had been the simple and easily fabricated circular shape and a non-optimal shape for maximum heat transfer by conduction, convection and radiation heat transfer. In this study, a topology based optimisation methodology was proposed for optimal shape selection from common shapes such as parabola, ellipse, rhombus, equilateral triangle and circular, using collection efficiency and collector element shape as the objective space and ease of manufacturability and cost as the decision space variables or constraints. The result of the research will help power generation and process engineers make decision on the best shape of receiver for process heat production for a minimal cost and ease of manufacture.

400564 Effect of Porous Media on the CO2 Hydrate Formation

Abstract:

The use of fossil fuels causes an increase in greenhouse gases, especially CO_2, in the atmosphere and is supposed to be a major factor on a global climate change. Many researches have tried to effectively reduce or to capture the major global warming gases. Especially, the hydrate formation mechanisms were intensively investigated for CO_2sequestration and storage process applications.

Gas hydrates are nonstoichiometric solid compounds of a polyhedral structure consisted of water molecules and interstitially encaging gas component molecules. The gas hydrates have a great potential to store and capture the vast amount of gas within them.

Higher solubility of hydrate forming guest in water and larger contact area between the hydrate formers and water is highly important. The contact area between the guest gas and water, which is the most critical factor for field operation, is dependent on the subcooling (ΔT) and driving force (ΔP) in the hydrate formation system. In recent years, researchers have suggested usage of certain additives to promote faster gas hydrate formation and increased surface area to volume of the fuel gas in hydrate forming systems. It has been shown that the gas hydrate formation could be faster in porous media and thus could be considered as kinetic promoter. The effect of different types of porous media, such as clays, sediment, activated carbon, carbon nanotube and silica gel, on hydrate formation are significant depending on the specific surface area, pore volume and pore size distribution.

Therefore, in this study, the effect of natural clay (montmorillonite and kaolinite) addition on the CO_2 separation and capture from fuel gas using the hydrate based gas separation process was evaluated. The CO_2 hydrate formation kinetics was investigated in the presence of porous media at constant temperature and pressure.

400652 Investigate of Polysilicon Chemical Vapor Deposition Reaction Kinetics Mechanism in Trichlorosiane-Hydrogen System

Abstract:

This study proposed a new surface kinetic model for polysilicon chemical vapor deposition reaction in trichlorosiane-hydrogen system. The mechanism taken into account the etching of hydrogen chloride by using the equilibrium constants calculated from the thermodynamic data consist of chemisorption of trichlorosilane and deposition by hydrogen. The reaction rate constants are evaluated by numerical simulation and comparison with experiments. The transport phenomena and reaction in CVD reactor has been described by using the surface kinetic model in conjunction with a detailed three-dimensional CFD transport model. The simulation results show that there is a good agreement between prediction of model and experiment. Furthermore, this surface kinetic model will be used for optimization of CVD reactors for epitaxial deposition of silicon.

401145 An Efficient Anode for Lithium Ion Battery: Inverse Silicon Opal Prepared By Magnesium Thermal Reduction of SiO2

Abstract:

Lithium ion batteries (LIBs) with high power performance and energy density are required for the substantial market growth in renewable energy storage and EV applications. Unfortunately, poor characteristics of commercial graphite; low theoretical capacity and poor rate capability, make it difficult to apply to high performance LIB anodes.

Several researches have reported transition metal oxides and alloys materials as high performance materials. Silicon has received much attention by its highly reversible capacity with lithium ions (4200 mAh g^{-1}). A crucial drawback of silicon, however, is to undergo huge volume change (> 300%), leading to mechanical stress of the electrode and pulverization of the particles during the alloying/dealloying with lithium ions. As a result, the rapid capacity fading occurs during cycling process. Many approaches have been studied to prevent the volume changes of silicon.

In recent years, the porous Si-based anode materials have attracted attention to avoid the common problems of Si powders such as volume expansion, electrically disconnected smaller particles and application. Especially, macroporous nano-materials prepared in the form of inverse-opal films were provided a number of advantages when used as anode/cathode active materials for LIBs. First, macroporosity can accommodate the volume change during the cycling process, resulting in maintaining the structure of electrode without collapse. Second, electrolyte and lithium ion can be easily penetrated and moved in the macroporos, leading to the improvement of rate capability and the decrease in the polarization.

In this work, an inverse opal structured-silicon will be fabricated by a magnesium reduction method. The inverse opal structure will be observed through X-ray diffractometer (XRD), scanning electron microscopy (SEM) and transmission electron microscopy (TEM). To evaluate the electrochemical characterization of the inverse opal structured-silicon, we will conduct coin-half cell tests including several electrochemical tests such as cyclic voltammetry (CV) and electrochemical impedance spectroscopy.

401344 The Tribology of Bicycle Chain Lubricants (Electronic Poster Presentation)

Abstract:

The Tribology of Bicycle Chain Lubricants

Properly lubricating a bicycle chain is arguably the easiest and least expensive method of improving a bicycle's performance. Biodegradable or biobased lubricants are desirable on the market today due to their performance as a safe, ecological, and effective product. In this study we report on the friction and wear properties of commercial bicycle lubricants, natural base oils (e.g. canola, soy, etc.) and greases, as well as the effect of certain additives on these properties. A pin-on-disk tribometer was used to precisely measure the coefficient of friction and produce a wear track for each lubricant sample. The resulting wear track was measured using an interferometer and the amount of wear is reported as a volume loss of material after a one hour trial. A correlation was derived to relate the coefficient of friction of each lubricant to an associated power loss while riding a bicycle. This power loss is what directly affects a cyclists' performance and corresponds to a loss in speed while pedaling the drivetrain. Commercial lubricants were tested to determine a standard of performance currently available on the market. Samples containing biobased materials were then tested and compared to the performance standard set by these lubricants. Combining these results we explore possible correlations between friction, wear, viscosity, density, and individual lubricant components or additives. Ultimately, we aim to arrive at an optimal composition for a biobased lubricant that minimizes both the power loss and wear associated with a bicycle chain in motion.

400599 Lignin Pyrolysis with Sodium Formate for Stabilization of Pyrolytic Products

Abstract:

With the rapid commercialization of cellulosic ethanol in recent years, utilization of lignin byproduct to value-added products has received increasing attention. Depolymerizing lignin into phenolic monomers could provide valuable precursors for biofuels and other biobased products. Although fast pyrolysis process has the ability to thermal depolymerize lignin, it usually yields a very small amount of phenolic monomers and instead produces a large amount of phenolic oligomers that have limited applications. Repolymerization of reactive phenols could be responsible for the increasing molecular weight of pyrolysis products. Introducing hydrogen atoms during pyrolysis could promote stabilization of phenolic monomers thus prevents repolymerization of phenolic monomers to oligomers. In the present study, sodium formate is selected as a hydrogen donating agent and co-pyrolyzed with lignin in a micropyrolyzer. It is found that the presence of sodium formate promotes the production of simple or stable phenols such as phenol, syringle and ethylphenol, while reduces the yield of acetic acid. Deuterated sodium formate is also employed in the present study to investigate the mechanism of hydrogen transfer during pyrolysis.

Pyrolysis of Concentrated Pulping Liquors to Generate Highly Deoxygenated Organics for Bio-based Fuels and Chemicals Production

Jamie St. Pierre, *Paul Rudnicki, Sedat Beis, and Adriaan van Heiningen*
University of Maine Department of Chemical and Biological Engineering, Jenness Hall, Orono, ME 04469

Abstract

The development of robust processes and the improvement of existing processes that produce liquid transportation fuels and chemicals using sustainable feedstocks is a rapidly growing field due to concerns surrounding global warming and the desire to reduce usage of non-renewable resources. In the northeastern United States particularly, liquid fuels and chemicals produced from lignocellulosic materials such as woody biomass are attractive due to the abundance and renewable nature of the feedstock, its ability to fix carbon dioxide during its growth period, and decades of experience utilizing sustainable harvesting practices. Black liquor, the main by-product of kraft wood pulping, is rich in organic material, such as degraded lignin and hemicelluloses, as a result of delignification and alkaline peeling reactions, respectively. Due to its significant organic content and large-scale annual production of approximately 200 million dry tonnes globally of kraft black liquor solids, it has been of interest as a source of renewable carbon for the production of liquid fuels and chemicals for almost a century.

The current work focuses on pyrolysis of concentrated soda-anthraquinone (AQ) black liquor (40 wt% dry solids) to produce a significantly deoxygenated organic fraction while retaining a high percentage of the organic carbon in this phase. Sodium formate is added to black liquor at various charges to investigate the effect on organic product distribution, quality and yield. Formate is utilized in this work as an *in situ* source of reactive hydrogen as it has been shown in previous studies to aid in hydrodeoxygenation as well as in reducing the amount of organic carbon lost as char during pyrolysis. The results demonstrate that as the formate loading is increased the organic yield increases while the carbon sequestered in the char decreases. The product distribution and quality of the organic fraction, however, remain fairly constant irrespective of the formate loading. The main identified products in the pyrolysis oil are phenolic in nature with varying degrees of alkylation on the aromatic ring. Other products mainly consist of cyclic ketones with trace amounts of aromatic hydrocarbons and carboxylic acids. Organic yields of up to 37 wt% on black liquor organics and a reduction of the organic carbon lost as char (from 49 wt% to 24 wt%) can be realized with this process at 50 wt% sodium formate loading on black liquor dry solids. The organic fraction is significantly deoxygenated with an O:C ratio of 0.1 based on black liquor organics (initial O:C of black liquor organics of 0.4) and an increased higher heating value (HHV) of 35 MJ/kg from 11 MJ/kg for the black liquor dry solids. Product yields and distributions after pyrolysis of concentrated black liquor for the various sodium formate loadings will be presented.

*** Formatting Issue - Best Available Paper/Graphic ***

Design of Biomass Pyrolyzer-Multiple Effect Distillation System Components for Laboratory Testing

Yunhe Zhang[1], Ali Amiri[1], O. John Idowu[2], Catherine E. Brewer[1]

[1]Department of Chemical & Materials Engineering, New Mexico State University
[2]Department of Plant & Environmental Sciences, New Mexico State University

Abstract

Multiple effect distillation (MED) is one of the most reliable thermal water desalination technologies. MED uses only low pressure steam and a minimal amount of electricity to power pumps. The steam can be generated by renewable energy sources such as geothermal or biomass rather than fossil fuels. Our overall project is the development of a lab-scale biomass slow pyrolyzer-MED system in which the thermal and electrical energy needed to operate the MED unit comes from the biomass. In order to obtain data needed in the system model, we have constructed various lab-scale components to represent unit operations of the whole system. The slow pyrolysis unit operation is represented with a batch, 10L tube reactor designed to operate around at 400°C under a nitrogen environment with slow heating rates.

Introduction

Biomass pyrolysis is a thermochemical process that heats organic material to temperatures over 400 °C in an oxygen-free environment to convert low energy-density biomass into a high energy-density oil (~22 GJ m^{-3} or ~17 MJ kg^{-1}), biochar (~18 MJ kg^{-1}) and syngas (~6 MJ kg^{-1})[1,2].

Slow pyrolysis is mainly used to produce biochar from low-value biomass feedstocks like yard waste or industrial byproducts, but can also be used to generate energy[3]. For slow pyrolysis, the heating rate is usually below 100 K/min, the reaction temperature range is from 300 °C to 800 °C[4], and the residence time varies from hours to days. The yields of slow pyrolysis are roughly 35% biochar, 30% bio-oil, and 35% syngas by mass[5].

There are two general methods of desalination: membrane-based desalination and thermal desalination. MED is a type of thermal desalination and consists of several chambers, called effects, with each functioning as a heat exchanger. In each effect, salt water is evaporated by being sprayed onto a tube containing hot steam. The evaporated steam then flows into the tube of the next effect to evaporate more salt water. For a MED unit, heat is needed to provide hot water or steam in the tube of first effect. To prevent scaling, generally calcium sulfate deposition, MED should be operated below 70 °C[6] under partial vacuum conditions.

Material and Methods

In this study, we used our reactor, a batch, 10L tube reactor designed to operate around at 400°C under a nitrogen environment with slow heating rate, to produce biochars from locally-available

*** Formatting Issue - Best Available Paper/Graphic ***

biomass feedstocks: pecan orchard prunings, pecan shells, cotton gin waste, tumbleweed, and yard waste.

Pyrolysis

Besides producing biochar, the feedstock materials also served to test the performance of our pyrolyzer. We first ran a hot flow test (without biomass) and then a shakedown trial (with biomass).

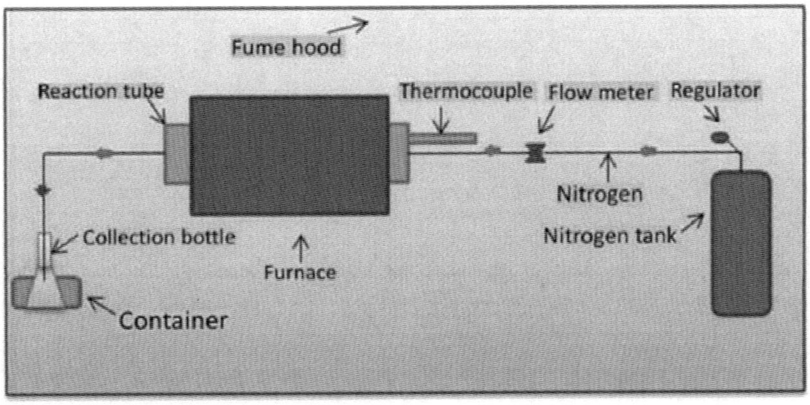

Figure 1. The lab-scale biomass slow pyrolysis reactor set-up.

The pyrolysis operation consisted of the following steps:

- Load the reaction tube with ~250g of biomass.
- Fill the condenser collecting bottle with ~700 ml of DI water, and then place it in an ice bath.
- Adjust the nitrogen flow rate to 1.0 L/min and the regulator's pressure to 2.75 bar.
- Set the gas outlet tubing heating tape's temperature to 200 °C.
- Program the furnace with a ramp rate of 5 °C/min, a target temperature of 450 °C and a dwell time of 60 minutes.
- Allow reactor to cool overnight under nitrogen flow once heating program has completed.

Results and Discussion

Hot Flow Test

Figure 2 shows that there was a temperature lag of about 22 °C between t temperature and the reactor temperature during the hot flow test. Since the reactor temperature does reach the furnace set temperature, we programmed our furnace to reach a maximum temperature 25°C above our desired reactortemperature.

*** Formatting Issue - Best Available Paper/Graphic ***

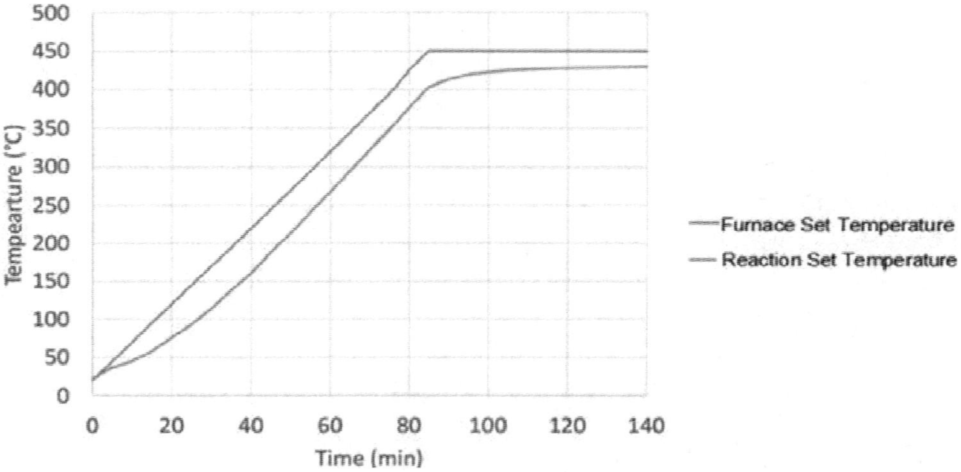

Figure 2. Hot flow test temperature vs. time graph showing reactor heating lag compared to furnace set temperature.

Shakedown Trial

Figure 3 shows the results of using 1 kg pecan trimming waste to perform a shakedown trial to test the reactor's mass and energy management. The shut down was required because our condenser was quickly overwhelmed as soon as the primary biomass decomposition reaction was reached. We concluded that the condenser system was not big enough for us to run at full 10-L biomass capacity. Instead, we would only use 250g of feedstock during our actual experimental trials.

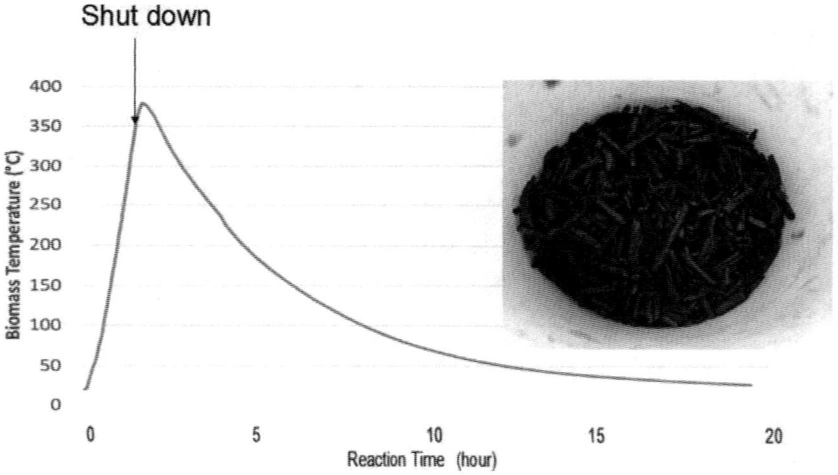

Figure 3. Shakedown trial temperature vs. time graphs showing reaction stopping point due to overwhelming the condensable vapor collection system.

Biochar Product

From the color, most of the feedstock was converted into biochar, but there remains a small amount of unreacted (brown) material. In the tumbleweed sample, a small amount of ash was also

*** Formatting Issue - Best Available Paper/Graphic ***

present, implying some combustion and therefore, some oxygen leakage into the reactor or, more likely, oxidation of the sample as the cooled biochar was removed from the reactor.

Figure 4. Biochar products

Conclusions

From our slow pyrolyzer experiments, we found that our pyrolysis reactor design has the advantages of temperature program control, controlled inert gas flow, (somewhat) increased feedstock capacity compared to other lab units, the ability to collect bio-oils, and ease of disassembly for cleaning. The disadvantages of our current system set-up are temperature limitations, capacity limitations due to the size of the ice-water bath condenser, space/length limitations in the fume hood, and small leaks from the reactor end caps.

Future Work

1. Characterization of biochar
2. Construction and shakedown of MED unit
3. Design of burner for pyrolysis vapors and non-condensable gases to replace condenser

References

1. Bridgwater, A. V., D. Meier, and D. Radlein. "An overview of fast pyrolysis of biomass." *Organic Geochemistry* 30.12 (1999): 1479-1493.
2. Ioannidou, O., et al. "Investigating the potential for energy, fuel, materials and chemicals production from corn residues (cobs and stalks) by non-catalytic and catalytic pyrolysis in two reactor configurations." *Renewable and sustainable energy reviews* 13.4 (2009): 750-762.
3. Downie, Adriana, et al. "Biochar as a geoengineering climate solution: Hazard identification and risk management." *Critical reviews in environmental science and technology* 42.3 (2012): 225-250.
4. Brown, Tristan R., Mark M. Wright, and Robert C. Brown. "Estimating profitability of two biochar production scenarios: slow pyrolysis vs fast pyrolysis." *Biofuels, Bioproducts and Biorefining* 5.1 (2011): 54-68.

*** Formatting Issue - Best Available Paper/Graphic ***

5. Goyal, H. B., Diptendu Seal, and R. C. Saxena. "Bio-fuels from thermochemical conversion of renewable resources: a review." *Renewable and sustainable energy reviews* 12.2 (2008): 504-517.
6. Alarcón-Padilla, Diego C., Lourdes García-Rodríguez, and Julián Blanco-Gálvez. "Design recommendations for a multi-effect distillation plant connected to a double-effect absorption heat pump: A solar desalination case study."*Desalination* 262.1 (2010): 11-14.

Pinch and Exergy Analyses within Process Simulation Software to Enhance Process Energy Efficiency

P. Baudet[a], O. Baudouin[b], S. Déchelotte[c], Q. Duval[d], and B. Wincure[e]

a ProSim SA, Immeuble Stratège A, 51 rue Ampère, F-31670 Labège, France, p.baudet@prosim.net,
b ProSim SA, Immeuble Stratège A, 51 rue Ampère, F-31670 Labège, France, o.baudouin@prosim.net
c ProSim SA, Immeuble Stratège A, 51 rue Ampère, F-31670 Labège, France, s.dechelotte@prosim.net
d ProSim SA, Immeuble Stratège A, 51 rue Ampère, F-31670 Labège, France, q.duval@prosim.net
e ProSim Inc, Science Center, 3711 Market Street, 8th Floor, Philadelphia, USA, b.wincure@prosim.net

Abstract:

Pinch analysis [1] is a well-known methodology, commonly used in the industry to optimize process energy consumption. Specific developments have been done in a commercially-available process simulator, ProSimPlus® Energy [2] to help process engineers in their daily work. Composite curves, characteristics of cold and hot streams of the process are automatically generated and an appropriate heat exchanger network is proposed by the software, depending on several constraints provided by the user. Energy management specific unit operations (heat pumps, ORC, boilers...) were also added in the library of the software to help the user to quickly analyze the impact on energy efficiency for this equipment. To tackle the challenge of process energy management enhancement, exergy analysis has been shown by Kotas [3] to also be a useful tool that exploits the concept of energy quality to quantify the portion of energy that can be practically recovered. Unfortunately, in contrast to enthalpy, this concept is less familiar to chemical engineers and can be rather difficult to handle. In particular, this physical quantity is rarely implemented in process simulators. To make exergy analysis more understandable and to demonstrate its value for the analysis of the energy efficiency of a process and its utilities, a fully-automated exergy analysis tool has been developed and integrated in ProSimPlus® Energy [3], [4]. The software combines these approaches, allowing the user to propose the most efficient process taking into account all of the defined constraints.

Keywords:

Energy efficiency, exergy analysis, ProSimPlus Energy, pinch analysis, simulation software.

1. Introduction

Process energy efficiency typically involves, among other things, optimization of the utilities production and methods used to reduce the process energy consumption through unit operations optimisation. Here we focus on two complementary methods: the Pinch method and the exergy analysis.

These two methods are implemented through process modelling carried out in the ProSimPlus process simulation software. The Pinch method and the exergy analysis are made possible by simply adding a totalizer unit operation into the process simulator, which then automatically performs the energy efficiency calculations.

The Pinch unit operation calls a fully-automated method that carries out the energy analysis (minimal requirement calculation for hot and cold utilities, calculation of the maximum quantity of recoverable energy) and that creates hot and cold streams, which become the input data for Simulis Pinch, a software that automatically creates exchanger networks.

The exergy analysis unit operation carries out the exergy balance for each unit operation of the process, as well as the global exergy balance. The exergy calculations include both the physical exergy, and also the chemical exergy.

These two methods are complementary, where the Pinch method is systematic, while the exergy method provides improvements when the Pinch method may fail, in particular for processes under pressure or for processes that can have a possible cascading of unit operations.

2. Process modelling

2.1. Collecting data

Data collection is a prerequisite for an accurate plant energetic optimisation: erroneous information on hot and cold streams will systematically lead to wrong process heat integration. Even if the heat integration methods can be considered as well-known, the fundamental risk is upstream, when streams characteristics are obtained, and data collection is the major step of the study. During this step, engineers have to identify streams which have to be cooled, heated, evaporated or condensed and have to extract, from all of the available information, only the characteristics which are key to the heat integration analysis. Streams with too much uncertainty on temperature or with fluctuating temperature have to be eliminated, hot and cold utilities have to be well identified and their consumption for each equipment has to be compared with the global production of the site. Engineers have also to pay attention to some potential source of errors: batch operations, heat integration network coupling production that involves different ways of operating, contractual restrictions, equipment technical bounds…

Process simulation is a very good way to improve the quality of the energetic optimisation and can be used to provide the engineer with missing information. The simulation model will also be used to test the heat integration network, change of operating conditions, process modifications, pinch value influence... Steady state process engineering simulation software are tools that perform mass and energy balance calculations for all the operating units of a continuous process in steady state operation. They include models based on first principles and allow the analysis of the behaviour of a complete process. The main functions of these software are the calculation of:

- all process stream characteristics (partial flowrates, temperature, pressure but also all of the physical properties);
- performance of the process equipment;
- all process data required for equipment sizing;
- data required for energy analysis of a process (pinch or exergy analysis).

Today, such software are widely used by process engineers in the design of new units as well as in operations of existing plants for process optimisation, units troubleshooting or debottlenecking, plants revamping, performing front-end engineering analysis… They are used in nearly all process industries: chemicals, pharmaceuticals, petrochemicals, oil and gas, refining, specialty chemicals… Even as similar unit operations (reactors, distillation or absorption columns, heat exchangers, pumps and turbines…) are commonly used in all of the industrial fields, material systems may have completely different behaviours. Thus, in addition to a complete unit operations library, process simulation tools have to embed thermo-physical calculation servers able to tackle the challenge of correctly representing these material systems.

2.2. Thermodynamic modelling

ProSimPlus® Energy includes the thermo-physical calculation server Simulis® Thermodynamics, which generates highly accurate pure component and mixture properties (thermodynamic, transport, compressibility…) and fluid phase equilibria. To obtain reliable thermodynamics calculations, one needs first to have accurate pure component properties. Based on the AIChE's DIPPR database [5], Simulis® Thermodynamics provides access to a property database of more than 2000 pure components. For each component, up to 125 constant properties (molecular weight, critical coordinates…) and 16 temperature dependant properties (vapor pressure, enthalpy of vaporization…) are available. All of these properties can be edited, plotted, modified and Simulis® Thermodynamics makes it very easy to create new components and private databases, offering pure component property estimation methods and experimental data fitting services. For multi-components systems, a major problem encountered in chemical engineering lies in the calculation of phase equilibria and thermodynamic properties (enthalpies, entropies…). Fluid phase equilibria can be calculated using two different methods. A first way of approaching the problem consists in applying to both phases different models: the liquid phase fugacities are calculated from a reference state, which is characterized by the pure component in the same conditions of physical state, temperature and pressure, where the laws of the ideal solutions are corrected by using a model of free excess enthalpy, also called an activity coefficient model (NRTL, UNIQUAC, UNIFAC…). The gas phase fugacities are calculated using an equation of state (ideal gas, SRK, PR…). These methods are used to represent the heterogeneousness of the considered system and are classically called "heterogeneous" methods. Their application is typically for the domain of low pressures and it is important to note that they do not satisfy the continuity existing in the critical zone between the vapor state and the liquid state. The second way of approaching the problem, grouping together the so-called "homogeneous" methods, applies the same model to both phases that are present. These "homogeneous" methods are generally an equation of state, in order to insure continuity at the critical point. For example, the equations of state with their classic mixing rules (SRK, PR, LKP…) belong to this second category. However, the domain of application of these models is limited to non-polar or few polar systems. By integrating the excess enthalpy models into the mixing rules of the equations of state, authors managed to combine these two approaches (MHV2, PSRK, NRTL-PR …). This type of model is usually called a "combined model", which is in fact a homogeneous approach with complex mixing rules. Moreover, some specific models were also developed for some domains of application: electrolytic mixtures, aqueous solutions of strong acid, formaldehyde… Simulis Thermodynamics is a unique software that includes all of the above methods and approaches.

3. Exergy analysis

3.1. Description

Exergy is based on simultaneous use of the first and second laws of thermodynamics and can be defined [6] as the maximum amount of work that can be obtained when a system is brought to thermodynamic equilibrium with the environment in terms of temperature, pressure and composition through reversible processes. For the calculation of the exergy of streams, the environmental state is chosen as twenty five Celsius degrees and one atmosphere. The standard chemical exergies of each compound tabulated by Rivero et al. [7] are used. For a given system, the exergy balance is given by the following equation, which, contrary to heat and mass balances, considers internal losses:

$$B_{in}^{matter} + B_{in}^{Heat} + B_{in}^{work} = B_{out}^{matter} + B_{out}^{Heat} + B_{out}^{work} + I$$

with:

$$B^{matter} = [h(T,P,z) - h(T^{00},P^{00},z)] - T^{00}[s(T,P,z) - s(T^{00},P^{00},z)] + \sum_{1}^{N_\varphi} \omega_\varphi \left(\sum_{i}^{N_c} z_i \left[b_i^{0\varphi} + RT^{00} \ln\left(a_i(T^{00},P^{00},z_i)\right)\right] \right)$$

$$B^{Heat} = Q.\left(1 - \frac{T^{00}}{T}\right)$$

$$B^{work} = W$$

The exergy balance can be refined by making the distinction between waste streams (directly rejected to the environment, i.e. external losses) and utilized streams (with a well-defined use, i.e. product streams). Both internal and external losses are sources of inefficiency in a process and one way to improve the energy efficiency of the system is to reduce these losses. Thus, a clear distinction between waste and used streams is required when performing the exergy analysis. From this exergy balance, several diagram types can be used and different performance indicators can be defined.

3.2. Implementation and use in ProSimPlus® Energy

The exergy analysis unit operation is simply added to the simulation process flowsheet, to indicate that the exergy balances will be calculated.

The user provides the data necessary for the analysis, such as:

- ✓ the balance zones (unit operations to take into account)
- ✓ the calculation type (physical exergy only, or chemical AND physical exergy)
- ✓ the loss streams of the process (to characterize the exergy losses)

The exergy bal;ance calculations are carried out at the same time as the classical material and energy balances of the ProSimPlus simulation.

In terms of results, the unit operation shows the exergy balances of each unit operation, using tables and charts, such as the Grassmann diagram below:

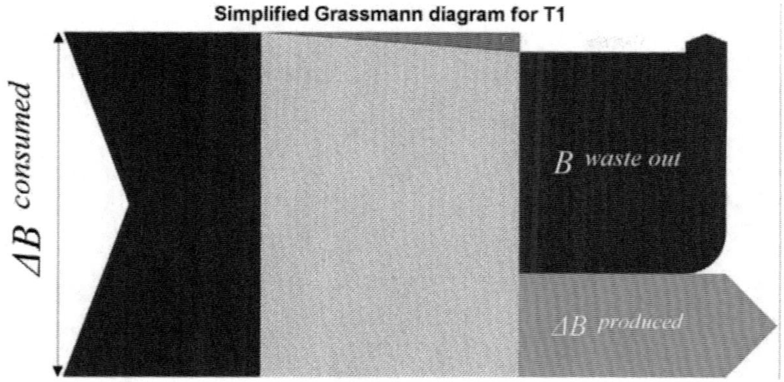

For each unit operation, but also for the whole process, exergy conversion balances appear in pie charts, as shown below:

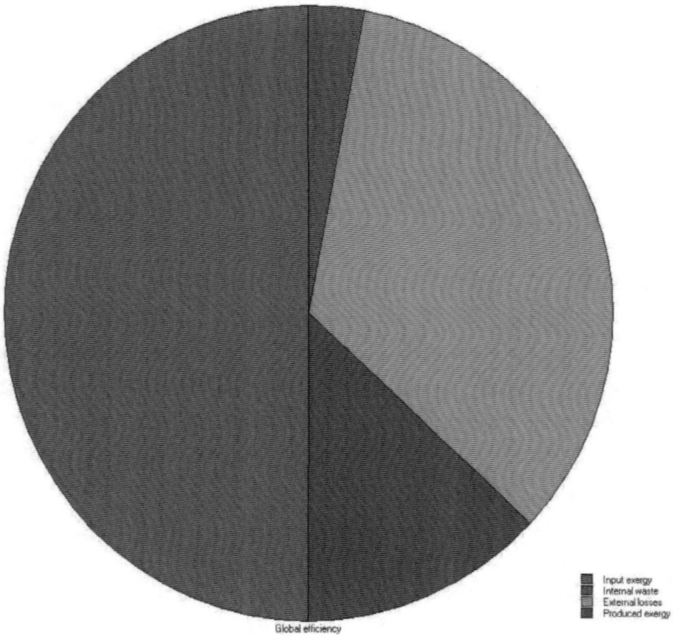

4. Pinch analysis

4.1. Diagnostic

Pinch analysis allows for the determination of the minimum required process utilities consumption. A dedicated pinch analysis unit operation is used in ProSimPlus to carry out energy integration calculations with the pinch method. Based on a chosen pinch that is acceptable for the whole process (minimum temperature difference between the hot streams and the cold streams), the minimum utility needs (hot and cold) are calculated as well as the maximum heat quantity recoverable by energy integration. The unit operation also plots the hot and cold composite curves, as well as the grand composite curve.

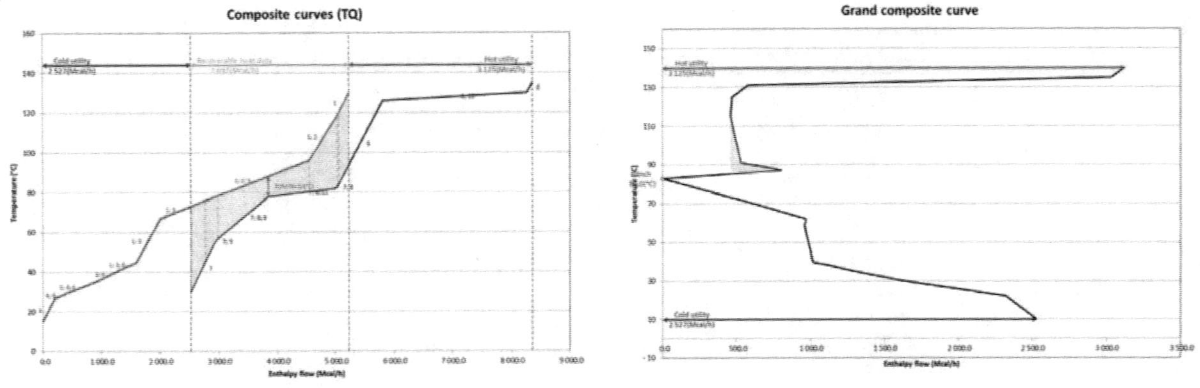

Fig. 1. Composite curves resulting from pinch analysis

Unit operations relevant to the pinch unit operation are those unit operations where a heating or a cooling occurs (boilers and columns condensers, heat exchangers, flashes, reactors...). To take maximum advantage of the pinch analysis unit operation, the following are important to consider during the modelling step:

- Mixers: according to the choice made by the modeller, the consequences of the heating and the cooling associated to the streams mixture can be, or not, taken into consideration. In particular, when streams with very different temperatures are mixed. In this case, the implicit cooling of the hottest streams and the implicit heating of the cool streams will automatically be taken into account.
- Pressure drop issues: the unit operations in which a pressure effect and a thermal effect can be met simultaneously must be used with care if a pinch study is planned. Indeed, antagonistic effects may appear and may lead to the creation of an outlier stream. For example, for a flash with a significant depression leading to an important temperature drop in the flash, with a slight heating, in order to reduce this temperature drop: from the unit operation point of view, heat is brought, yet the temperature decreases between the inlet and the outlet. The question is the following one: is the stream to take into consideration hot or cold? In such situations, the stream is not taken into consideration and a warning message appears. For the unit operations that give the option to define a pressure drop, this possibility may be used to describe the unit operation pressure drop, but not the pressure drop of a valve located upstream of the unit operation (in particular for the simple heat exchanger, or the Coolers/Heaters...). In the same way, for the flashes, it is advisable to represent the valve upstream of the flash that carries out the effective pressure drop.
- Existing heat exchangers: if one wants to keep the existing integration exchangers, it is possible to specify that the streams have been integrated, and then they will not be taken into account in the diagnostic.
- Process outlets: they can also be heat sources (hot effluents that may possibly be cooled) or on the opposite, cold sources. However, it is necessary to define a cooler/heater on theses streams to indicate if an energy recovery is possible or not on this stream.
- Phase change: if a stream changes partly or totally its physical state, the isobar heating capacity of this stream varies greatly along the exchanger. It is possible to linearize this variation in order to find a mean isobar heating capacity or to segment it according to the physical states. This segmentation makes it appear as if there are up to three different streams for the same stream. For instance, for a liquid vaporization, the three segments apply to:
    - The liquid heating to the bubble point;
    - The state change to the dew point;
    - The steam superheating.

The calculation can be done after the simulation of the whole process in order to simply achieve an energy balance on the process. For a more sophisticated analysis, it can also be carried out in "On-Run" mode in order to do case studies on some process parameters or to use the pinch unit operation results as a criterion or constraint for an optimisation issue.

4.2. Heat exchanger network

The energy integration concept is used to arrive at the optimal synthesis of a heat exchanger network that achieves the maximum energy recovery, i.e. to reach the process operating point with the minimum amount of required energy. This synthesis is subject to construction rules (division of the problem into two sub-problems, one below and the other above the pinch temperature). The ideal point may not be simple to reach (due to partial exchanges on streams, stream divisions...), and the number of heat exchangers needed may be significant. Optimisation tools can help to determine the optimal structure of the heat exchanger networks using MINLP-type algorithms. Even though these tools are powerful, their calculation time can be very long, and their convergence on an industrial scale is not guaranteed.

The approach developed by Simulis Pinch is more pragmatic: the aim is to propose very quickly the right solutions with a limited number of heat exchangers. The method is not an "optimal" method in the mathematical sense of the term, as it systematically aims to exchange the maximum power between two streams. If a hot stream and a cold stream can exchange heat, the algorithm searches for the possible couplings allowing the exchange of the maximal thermal power under certain limitations or "boundary conditions". Between two streams (cold stream and hot stream) that can exchange heat, the combination of the possible exchanges is reduced or limited (maximum of 3 possible couplings between a hot and a cold stream). For a problem with HSN hot streams and CSN cold streams, the number of exchanges is then less than or equal to 3*HSN*CSN, where HSN is the number of hot streams and CSN is the number of cold streams.

The enumeration phase consists in determining the exact number of possible exchanges between the matched hot and cold streams. Each possible exchange is characterized by a certain number of attributes, e.g. the power exchanged, the efficiency...

Efficiency for the exchange is defined by the following formula:

$$Eff = \frac{MRE\_AF + P}{MRE\_BE}$$

where:

P	power exchanged by the coupling considered
MRE_BE	Maximum amount of Recovery Energy BEfore the coupling is considered
MRE_AF	Maximum amount of Recoverable Energy AFter considering the coupling, i.e., it is the MRE obtained by a diagnostic operation by updating the streams with the assumption that the coupling has been carried out.

It has to be noticed that if all the couplings selected have an efficiency equal to 1, the approach is by definition optimal. In other words, the choice of an exchange does not deteriorate the initial recovery potential.

The Simulis Pinch operation algorithm for the synthesis of the heat exchanger network is described in Fig. 2.

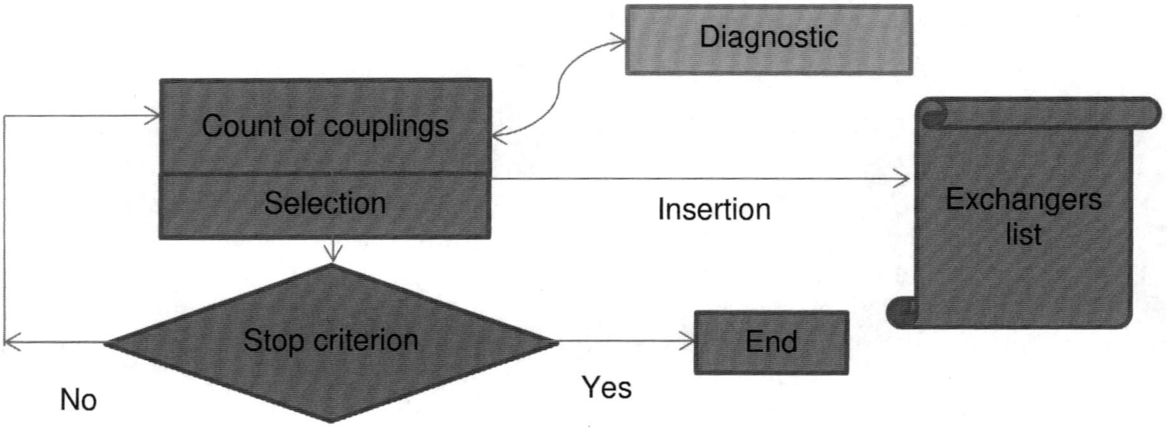

Fig. 2. Simulis Pinch operation algorithm

The count method provides an exhaustive list of the possible exchanges. In addition to the thermodynamic criteria, one can specify constraints that will allow only the pertinent exchanges, e.g. minimal power exchanged, a minimum recovery percentage (compared to the initial MRE)…

The selection method consists in selecting an exchanger among all the possible exchanges. Several selection criteria are possible, e.g. maximal heat duty exchanged, maximal efficiency, HeatDuty*MaximalEfficiency… One can also use a manual method and select the heat exchangers at each iteration.

Finally, different stop criterion/criteria can be selected, e.g. maximum number of exchangers reached, recovery percentage reached (% compared to the initial MRE)…

The list of proposed exchangers is the main result of the synthesis and all the selected exchangers are listed with the required information needed to estimate their cost.

5. Specific unit operations and economic evaluation

With either the Pinch method or the exergy analysis, the investigations may lead to a consideration of the implementation of specific unit operations to optimize the energy integration of the process.

Obviously, the time of return on investment cannot be compared with the heat exchangers proposed by the automatic synthesis of the heat exchangers networks coming from the Pinch analysis. Nonetheless, in some cases, the energy gain associated with the implementation of a thermal unit operation is such that the time of return on investment may be reasonable and worth considering.

In order to test the insertion of such complex unit operations, specific macro-unit operations have been developed in ProSimPlus:

- Simple heat exchanger
- Two-stage or side-injection heat pump
- Rankine cycle
- Gas engines
- Gas turbines

These macro-unit operations are easily inserted into the simulation process flowsheet. An example of the heat pump unit operation is shown below:

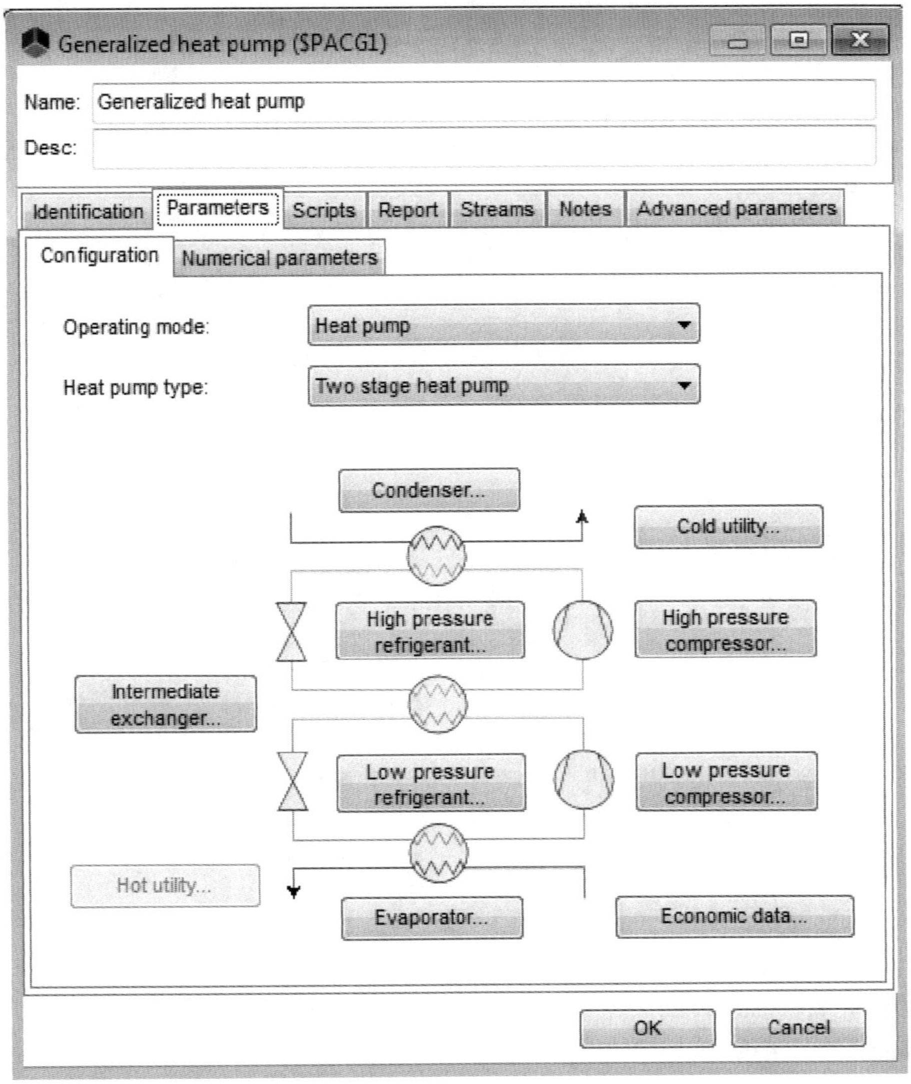

In order to test the economic interest of such an investment, a financial balance unit operation is also available in the latest version of ProSimPlus. This unit operation is also a totalizer unit operation that provides a full economic balance of the process (operating, maintenance and investment costs). This unit operation can calculate the time of return on investment related to the addition of a unit operation such as a thermal unit operation.

6. Perspectives and conclusions

The two energy balances described above (Pinch analysis and exergy balance) are now easily accessible in the ProSimPlus simulation environment. Engineers at all levels (both senior and junior!), can now use these tools to easily carry out expert analysis and to propose energy improvements for their processes.

Prospects for an improvement are of two kinds:

The exergy analysis provides a focus on the irreversibilities of the process, but does not yet provide concrete proposals to the user. The work that remains consists in devising methods to guide the user to formal solutions (in an automated way), or at least, to bring attention to the ideas that are most important to consider for future investigation.

✓ Both methods exist in ProSimPlus, but it would make more sense to combine them in a global approach, in order to fully systematize the energy efficiency of the process. Work is underway to implement such a methodology.

To conclude, the work performed helped to achieve a significant headway in the areas of decision support and automatic realization of the energy efficiency, from a simple simulation of the process. The ideal method that combines both approaches in a systematic but fully automated approach still remains to be formalized.

Acknowledgments

Financial support provided by the French National Research Agency (ANR) is gratefully acknowledged (convention n° ANR-11-SEED-0012-03). ANR-COOPERE-2 project (Combiner Optimisation des ProcédEs, Récupération énergétique et analyse Exergétique - COmbining Process optimisation, Energy Recovery and Exergy analysis) involved several partners (Veolia Environnement Recherche et Innovation, AgroParisTech, ProSim S.A and Laboratoire de Génie Chimique) and aimed at integrating in the process simulation software ProSimPlus® a complete methodology of process energy analysis.

Nomenclature

a	molar activity
B	exergy flow, kW
b	molar exergy, kJ/mol
h	molar enthalpy, kJ/mol
I	irreversibility or internal losses, kW
Nc	number of components
$N\varphi$	number of phases of a stream
P	absolute pressure, atm
R	universal gas constant, kJ/mol K
s	molar entropy, kJ/mol K
T	absolute temperature, K
W	power, kW
z	global molar composition of a material stream

Greek symbols

ω	molar vapour fraction of a stream

Subscripts and superscripts

00	environmental state
φ	related to a phase

References

[1] Linnhoff B., Thermodynamic Analysis in the Design of Process Networks [dissertation]. Leeds, UK: Leeds University; 1979

[2] ProSim web site – Available at:< http://www.prosim.net/en/index.php> [accessed 20.01.2015].

[3] Kotas T. J., The exergy method of thermal plant analysis. Butterworths; 1985

[4] Ghannadzadeh A., Thery-Hetreux R., Baudouin O., Baudet P., Floquet P., Joulia X., General methodology for exergy balance in ProSimPlus® process simulator, Energy 2012; 44: 38-59

[5] Rowley R. L., Wilding W. V., Oscarson J. L., Yang Y., Giles N. F., DIPPR® Data Compilation of Pure Chemical Properties, Design Institute for Physical Properties, AIChE (2011)

[6] Szargut J.,Morris D.R., Stewart F.R., "Exergy Analaysis of Thermal Chemical and Metallurgical Processes". New York: Hemisphere Publishing Corporation; 1988

[7] Rivero R., Garfias M., "Standard chemical exergy of elements updated", Energy 2006; 31(15): 3310-3326

Operational optimisation of industrial cooling water systems

Fei Song[(1)], Yichong Wang[(2)], Nan Zhang, Robin Smith,

(1) Centre for Process Integration, School of Chemical Engineering and Analytical Science, The University of Manchester, M13 9PL, UK, fei.song-2@postgrad.manchester.ac.uk

(2) No.217 Jiefang Road, Zhongshan District, Dalian, 116001, China, Tel: +8613941193800

Cooling water systems are used to remove wasting heat from processes, in which cooling towers are used to cool hot water from cooler networks by heat and mass transfer between bulk water and air to produce cold water for cooler network, and water pumps are employed to provide the pressure for water transportation. Cooling water systems consume a large amount of water and energy in chemical plants. The investigation on cooling water systems is important in water and energy conservation of the process industry. The key components, namely cooling towers, coolers network and pumps, interact with each other in cooling water systems. Therefore, minimum water and energy consumption in a cooling water system should be determined by considering these components simultaneously.

Research [1,2,3] on operational optimisation of cooling water systems to date has considered the components in cooling water systems simultaneously when determined operating conditions. However, there are some limitations in the previous work. A simple and less accurate cooling tower model was employed in cooling water system modelling. In addition, the previous work has focused on simple cooling water systems with coolers in parallel configuration and a single cooling tower, instead of considering the combination of series and parallel cooler configurations and multiple cooling towers.

In this work, a mathematic model of overall cooling water systems including cooling towers, cooler networks and pumps is developed, in which a simplified cooling tower model is regressed out based on a rigorous cooling tower model[4,5] for forced draft counter flow wet cooling towers. The cooling tower model is validated to provide reliable results of evaporation rate and water outlet temperature of cooling towers. Cooler networks are modelled for the combination of parallel and series configurations. As the model of cooling water systems is formulated as an NLP problem, CONOPT solver is then applied for optimisation of relevant operating conditions of cooling towers and water coolers. A case study is presented to illustrate the application of the model developed in this work for cooling water systems: multiple cooling towers and a cooler network in series and parallel arrangement. After optimisation, circulating water flowrate in the case is reduced by around 9.5%, make-up water is reduced by about 12.6%, and the total operating cost is decreased by around 6.7%. The results show that higher inlet water temperature of cooling towers with lower circulating water flowrate tends to reduce the operating cost.

References:

[1] M.M. Castro, T.W. Song, J.M. Pinto, Minimization of operational cost in cooling water systems, Trans IChemE 78(Part A) (2000), 192–201

[2] G.F. Cortinovis, M.T. Ribeiro, J.L. Paiva, T.W. Song, J.M. Pinto, Integrated analysis of cooling water systems: Modeling and experimental validation, Applied Thermal Engineering (2009).

[3]G.F. Cortinovis, J.L. Paiva, T.W. Song, J.M. Pinto, A systemic approach for optimal cooling tower operation, Energy Conversion and Management 50 (2009), 2200–2209

[4] M. Poppe, H. Rogener, Berechnung von Ruckkuhlwerken, VDI-Warmeatlas (1991), Mi 1–Mi 15

[5] J.C. Kloppers, D.G. Kroger, A critical investigation into the heat and mass transfer analysis of counterflow wet-cooling towers, International Journal of Heat and Mass Transfer 48 (2005), 765–777

401199 Performance Simulation and Optimization of IGCC Power Plants

Abstract:

Performance simulation and optimization of Integrated Gasification and Combined Cycle (IGCC) power plants

Chengjun Qian and Nan Zhang

Centre for Process Integration

School of Chemical Engineering and Analytical Science

University of Manchester, UK

M13 9PL

chengjun.qian@postgrad.manchester.ac.uk

Integrated Gasification and Combined Cycle (IGCC) has arisen as a promising technology in electricity production due to its higher efficiency, lower emitting level, capability for CO_2 capture and flexibility in feeds and products. However IGCC is yet to be competitive against conventional technologies due to its higher capital and operation costs, lower reliability and under-utilised CCS ability. Much research has been done to improve its thermal efficiency, economic and environmental performance. As IGCC is a combination of two well-developed technologies - gasification and combined cycles, some of the key potentials for improvement need to be exploited through their integration.

In this work, a general framework is proposed to incorporate property calculation, unit operation models, and flowsheet synthesis in an overall systematic model. Rigorous property methods and relevant parameters for gas streams, steam and electrolyte solutions come from well-developed methods and databases, such as PR-BM method for gas, IAPWS-IF97 for steam, and E-NRTL for sour gas treatment(Mathias 1983; Wagner and Kretzschmar 2007; Zhang and Chen 2010) . These rigorous methods are able to give more reliable results than approximate ones, which apply linear approximations for heat capacity, enthalpy or reaction constant calculation. Unit models are grouped according to operations involved and modelled based on physical laws and design specifications. Part load performance curves for key units, such as gasifiers and gas turbines, are included to facilitate flow sheet analysis. The units are linked to each other through connecting streams to form a structured flow sheet.

In general the synthesized flowsheet is formulated as an MINLP problem, which is difficult for most solvers. Therefore, the MINLP problem is degraded to NLP sub-problems combined with enumeration of discrete decision variables to remove integer variables. The resulted equation based model is optimised with the CONOPT solver in GAMS™ software. Key decisions such as feed stock selection, product yields and operating parameters can be optimised simultaneously. However the NLP sub-problems are still difficult to solve and no global optimum is guaranteed. Several technologies, such as simulation-based initialization, random start points and shortcut models, are considered to help convergence and improve solution quality.

A case study is carried out, and results show that major performance improvement can be achieved towards affordable IGCC power plants.

Mathias, P. M. (1983). "A versatile phase equilibrium equation of state." Industrial & Engineering Chemistry Process Design and Development **22**(3): 385-391.

Wagner, W. and H.-J. Kretzschmar (2007). International Steam Tables-Properties of Water and Steam based on the Industrial Formulation IAPWS-IF97: Tables, Algorithms, Diagrams, and CD-ROM Electronic Steam Tables-All of the equations of IAPWS-IF97 including a complete set of supplementary backward equations for fast calculations of heat cycles, boilers, and steam turbines, Springer.

Zhang, Y. and C.-C. Chen (2010). "Thermodynamic modeling for CO2 absorption in aqueous MDEA solution with electrolyte NRTL model." Industrial & Engineering Chemistry Research **50**(1): 163-175.

Aspen Plus simulation of biomass gasification in a downdraft gasifier

Jia Yu, Joseph D. Smith, PhD

Missouri University of Science and Technology, Rolla, Missouri, USA

EXTENDED ABSTRACT

Biomass is widely perceived as a renewable energy resource which can be used to produce fuels, chemicals and power. Biomass gasification is an important method to obtain syngas from biomass feedstock. A significant amount of experimental data is available and many models have been developed using a wide variety of computational tools to analyze downdraft biomass gasification which are discussed in the literature. Aspen Plus modeling relies on minimization of Gibbs (APMG) free energy to approximate the chemical equilibrium occurring in a biomass reactor. This methodology has been used to simulate the conversion of biomass to final products. This approach has been applied for selected values of equivalence ratio (ER) due to a lack of information on reaction pathways, reactor profile and operating conditions.

In this paper, a detailed Aspen Plus model is developed and applied to a downdraft biomass gasifier is described. The gasifier model uses four "Plug Flow Reactor" (PFR) blocks in sequence to approximate the homogeneous gas phase reaction occurring in the downdraft gasifier with different solid and gas residence times. The PFR blocks are also combined with a "Continuously Stirred Reactor" (CSTR) block to approximate the heterogeneous reactions occurring in the biomass as it moves through the gasifier. Gasification kinetic reactions identified in the literature were inserted into the reactor blocks using an explicit methodology. The model predicts syngas composition over a wide range of ER to match measure peak of CO Fraction in the syngas product. The model was then used to optimize the selected operating conditions for the downdraft gasifier. These predictions were compared with both experimental data from the literature and APMG model predictions.

401713 Modelling of a Solar-Thermal Reactor for Biomass Fast Pyrolysis

Abstract:

Solar thermochemical conversion can provide carbon-free processes for biomass to fuels and chemicals. Biomass fast pyrolysis is an endothermic process that requires fast heating rates (>300 ^{o}C) and high heat flux at temperatures up to 600 ^{o}C. Concentrated solar thermal energy is an ideal option to meet these requirements. A computational Eulerian-Eulerian model is used to simulate and investigate fast pyrolysis of switch grass biomass in a solar parabolic trough receiver/reactor equipped with a novel gas-separation system. The separator controls the effect of tar-cracking reactions and achieves high separation efficiency compared to other gas-solid separation methods such as the U-separator and the cyclone. The study assumes an average heat flux concentrated along the receiver/reactor. Pyrolysis reaction was represented as a single global first order Arrhenius type reaction with volatiles separated into condensable (bio-oil) and non-condensable gas (NCG) products. The drying of moisture of the switch grass was represented as a mass transfer process in which the liquid water is converted to vapour.

The mechanism of solid-gas separation, residence time, devolatilization efficiency and product yield has been presented in this model. The separation efficiency achieved by the conical deflector was over 99%, this was found to correlate with experimental separator conducted by Huard et. al. (2010). The gas residence time distribution was within 1-2 s; this is within the range of residence time recommended for fast pyrolysis. The product composition had a bio-oil yield of >50% with minimal NCG. The range of yields was compared with similar models and experiments using different reactors. This work presents a new and energy efficient approach to demonstrate the potentials of solar thermal systems for biomass fast pyrolysis.

401220 Mixing Simulations for the Scaling-up of Succinic Acid Production from Biorefinery Glycerol

Abstract:

The main objective of this work is to investigate the hydrodynamics in a biochemical process at different scales to improve its performance. The scaling-up of a bioprocess from a small scale to a large scale reduces the main product yield mostly due to the alteration of the mixing behaviour in different reactor sizes. It is observed that the mixing efficiency is reduced with increasing reactor size. Scaling-up of biochemical systems results in insufficient gas supply, noncontrolled pH and substrate gradients, due to unsatisfactory mixing (Oosterhuis and Kossen, 1984; Larsson et al., 1996). Furthermore, reaction times are often shorter than mixing times in lab-scale reactors in contrast with industrial scale reactors where mixing times are much larger (Akiti and Armenante, 2004).

Recently, Theodoropoulos and co-workers (Vlysidis et al., 2011; Rigaki et al., 2013) have developed a (fed)batch reactor model for the prediction of succinic acid (SA) production from glycerol in order to calculate the substrate, product and byproduct(s) consumption/production rates. The constructed ordinary differential equation (ODE) based model was deemed to accurately predict the concentration transients in small-scale reactors. Nevertheless, it was observed that scaling-up resulted in differences between the model predictions and the experimental observations, mainly due to differences between the small and the large bioreactors' principal hydrodynamics. For this purpose, this study focuses on the behaviour of macro-mixing in the scaling-up of such a stirred tank bioreactor system. Macromixing accounts for the flow processes manipulating the mean concentration and the residence time needed for the molecules to flow through the reactor (Vicum et al., 2004).

Computational fluid dynamics (CFD) simulations are performed here, employing COMSOL Multiphysics, to investigate the mixing importance in such systems and to complement the system parameters' optimisation as well as the robust system design and control. The proposed CFD model couples the k-ε model used for the simulation of turbulent kinetic energy (k) and of energy dissipation rate (ε), with convection/diffusion as well as the reaction processes taking place in the fermentation system. The resulting set of partial differential equations is solved simultaneously leading to efficient prediction of the concentration transients in the volume of the agitated fed-batch bioreactor.

References

Akiti O. and P.M. Armenante, AIChE Journal, 50:566-577, 2004.

Larsson G., M. Tornkvist, E.S. Wernersson, C. Tragardh, H, Noorman and S.O. Enfors, Bioprocess and Biosystems Engineering, 14:281-289, 1996.

Oosterhuis N.M.G. and N.W.F. Kossen, Biotechnology and Bioengineering, 26:546-550, 1984.

Rigaki A., C. Webb and C. Theodoropoulos, Chemical Engineering Transactions, 35:1033-1038, 2013.

Vicum L., S. Ottiger, M. Mazzotti, L. Makowski and J. Baldyga, Chemical Engineering Science, 59:1767-1781, 2004.

Vlysidis A., M. Binns, C. Webb and C. Theodoropoulos, Biochemical Engineering Journal, 58-59:1-11, 2011.

399510 Reduce Energy Consumption in Utility Centers with Real-Time Dispatch Optimization

Abstract:

While or because natural gas pricing is at historic lows, electrical energy costs are getting more attention. Data centers are the fastest growing industry sector in energy consumption and alone consume a measurable fraction of total US generated electricity. Semiconductor manufacturing, like many peers (automotive paint shops…) have tight environmental specifications that drive utility costs to be a significant fraction of production costs. First-stage energy reduction solutions have been accomplished and there is increasing interest in an intelligent control system that can dispatch chillers, compressors, thermal storage, turbines or combined cycle units as efficiently as possible and to do this automatically. Real-time optimization or dispatch optimization of available equipment is designed to drive unit selection and unit loading to the best performance curve balance based on current conditions and demands to minimal energy costs.

The hurdle with traditional solutions has been that accurate models needed for ideal selections are difficult to sustain. But modern software environments have continued to make this easier than ever so that a decision hurdle may be now overcome. A strategy will be presented to reduce installation and maintenance costs and efforts leveraging graphical optimization with a robust MINLP solver.

399783 Design Optimisation of Multi-Stream Plate Fin Heat Exchangers

Abstract:

Design Optimisation of Multi-stream Plate Fin Heat Exchangers

Kunpeng Guo, Nan Zhang, Robin Smith

Centre for Process Integration, School of Chemical Engineering and Analytical Science, the University of Manchester, Manchester, M13 9PL, United Kingdom

With small minimum temperature approach, large surface area per volume, high heat transfer coefficient and possibility of handling several streams, multi-stream plate fin heat exchangers, categorized as compact heat exchangers, are widely used in various industrial areas such as chemical, petrochemical processes, cryogenics and aerospace industries [1]. Compared with shell and tube heat exchangers, plate fin heat exchangers have larger heat transfer load per unit volume. Consequently, the capital cost and the process operating cost will be reduced to a large extent.

However, according to Kays and London [2], there are approximately sixty standardised fin parts with fixed geometry and heat transfer and pressure drop performance. A major challenge in designing optimal multi-stream plate fin heat exchangers is the large number of combinations of standardised fin geometries for various fin types to choose from, which adds discrete aspects to an already complicated design problem. Another challenge is the difficulty of simultaneous optimisation of fin selection and imposed pressure drop constraints caused by discrete heat transfer and pressure drop performance of standardised fin types [3], especially if one wants to consider the possibility of mixing different fin types in the same device to enhance the heat transfer performance.

In this work, a new design algorithm is proposed to mix and match fin types used in a single heat exchanger and optimise fin selection and consider pressure drop simultaneously. By treating basic fin geometries such as plate spacing, fin pitch, fin thickness and fin length as variables, and considering thermal hydraulic performance of different fin types as a function of basic fin geometries, the optimal fin types and their corresponding design parameters can be obtained by minimising the total volume or capital cost of heat exchanger. A case study from the literature are presented to illustrate the efficiency and accuracy of the proposed algorithm. The result shows that the new algorithm can give an optimum configuration with short computational time and accuracy compared to the published design methodologies.

Key words:

Plate-fin heat exchanger design optimisation mix and match

Reference:

[1] Yousefi, M., Enayatifar, R., Darus, A.N. and Abdullah, A.H., Optimization of plate-fin heat exchangers by an improved harmony search algorithm. Applied Thermal Engineering, 50 (2013): 877-885.

[2] Kays, W. M. and London, A. L., Compact heat exchangers. New York: McGraw-Hill Book Co, 1984.

[3] Guo, K., Zhang, N., Smith, R., Optimisation of fin selection and thermal design of plate-fin heat exchangers, Chemical Engineering Transactions, 39(2014), 325-330.

401319 A Simple Model to Predict the Pressure Drop in Three Phase Inverse Fluidized Bed

Abstract:

A simple model to predict the pressure drop in three phase inverse fluidized bed

R. Trivedi*, T. Renganathan, K. Krishnaiah

Department of Chemical Engg., IIT Madras – 600036, India

Abstract

Three phase inverse fluidized beds are contacting patterns in which continuous downflowing liquid phase and upflowing dispersed gas phase maintain a bed of suspended particles with density lower than the liquid. Compared to classical cocurrent three phase fluidization, higher mass transfer rates are achieved in inverse fluidized beds, due to the low inertia of particles used in the system. Owing to this advantage IFBs are used in wastewater treatment and biochemical processes. Successful design and operation of IFB requires information on hydrodynamics, transport rates, kinetics and contacting. Hydrodynamic characteristics include flow regime, pressure drop and phase holdups. Several studies have been conducted on the hydrodynamics of three phase IFB [1, 2]. However, no attempt has been made to predict pressure drop in terms of operating parameters. Hence the objective of the present work is to predict pressure drop in three phase IFB using gas holdup in bubble column and weight of particles.

Experiments are conducted for bubble columns and three phase IFB in an acrylic column of 89 mm I.D. and 186 cm height. Measured flowrates of water and air are allowed countercurrently into the column through distributors at the top and bottom respectively. Spherical polypropylene particles are used as solid phase. L-tube manometers and conductivity probes were used to measure pressure drop and holdups respectively. Experiments were conducted in bubble column for different liquid and gas flowrates. Data were taken on the height of gas liquid dispersion and pressure drop using which gas holdup was calculated. Experiments were conducted in IFB with particles at the same gas and liquid flowrates as those used for bubble column experiments. Data were taken on pressure drop, bed height, and conductivity.

In bubble column, the pressure drop and the gas holdup increase with gas flowrate. While homogeneous bubbling regime was observed at low gas flowrates, coalescence of bubbles occurred at higher gas flowrates reducing the rate of increase of pressure drop and gas holdup. Liquid flowrate has a marginal effect on pressure drop and gas holdup in the countercurrent bubble column. In three phase IFB experiments, it is observed that for particles of given characteristics and bed height, pressure drop increases with increase in gas and liquid flowrates.

It is hypothesized in the present work that pressure drop in three phase IFB at a particular gas and liquid velocities can be considered to be the sum of pressure drop in bubble column at the same velocities and the net buoyant force of the bed of particles per unit cross-sectional area of the column, assuming pseudo fluid (gas liquid mixture) [3]. Accordingly, the pressure drop in IFB is predicted by adding the pressure drop measured in bubble column and net buoyant force per unit area of the bed of particles. The predicted values compared satisfactorily with the experimentally measured pressure drop in IFB with an RMS error of 9%.

To make the calculation of pressure drop in IFB completely predictive, the pressure drop in bubble column was correlated by proposing a new empirical equation for gas holdup in countercurrent bubble column [4]. Using the predicted value of pressure drop for bubble column instead of experimental value, the pressure drop in three phase IFB could be predicted with an RMS error of 11%. Given the complex interactions between the phases and the simplicity of the model, the error seems to be satisfactory.

References:

1. Krishnaiah, K., S. Guru and V. Sekar, "Hydrodynamic Studies on Inverse Gas-liquid-solid Fluidization", *Chem. Eng. J.,* **51**, 109-112 (1993).

2. Renganathan, T. and K. Krishnaiah, "Prediction of Minimum Fluidization Velocity in 2- and 3- phase Inverse Fluidized Beds", *Can. J. Chem. Eng.,* **81**, 853-860 (2003).

3. Felice R. D., "The pseudo-fluid model applied to three-phase fluidisation," *Chem. Eng. Sci.,* **55**, 3899-3906 (2000).

4. Koide K., S Morooka, K. Ueyama and A. Matsuura, "Behaviour of bubble swarm in large scale bubble column," *J. Chem. Eng. Jpn*., **12**, 98-104 (1979).

401742 New Mesoporous ZSM-5 Synthesis and Its Use in Bio-Oil Hydrodeoxygenation

Abstract:

Hydrodeoxygenation has been considered to be one of the most promising methods for bio-oil upgrading. Here, the catalytic performance of a newly developed hydrogenation catalyst, platinum supported on mesoporous ZSM-5 (Pt/MZSM-5), was tested in model bio-oil compounds hydrodeoxygenation. Pt/ZSM-5 and Pt/Al_2O_3 were used as reference. Three common bio-oil compounds with various oxygen contents and molecules sizes, guaiacol(GUA), cresol, and dibenzofuran(DBF), were used as model bio-oil compounds to evaluate the bio-oil hydrodeoxygenation on these three catalysts. Our results indicate that Pt supported on mesoporous ZSM-5 show better performance in bio-oil hydrodeoxygenation than Pt/ZSM-5 and Pt/Al_2O_3. The excellent catalytic performance is attributed to the combination of strong acidity and mesopore structure in mesoporous zeolite.

400084 The Hydrogenation of Simulated Catalytically Upgraded Biomass Pyrolysis Oils

Abstract:

Biomass pyrolysis oils can be produced by rapidly heating biomass materials such as wood, corn stover, or straw to500°C for approximately 1 s and then cooling rapidly. The resulting liquids are combustible, but have many undesirable properties as fuels. Therefore, efforts are underway to upgrade these liquids to better fuels.

One way of improving these oils is to catalytically upgrade them during the pyrolysis process to reduce the oxygen content and then hydrogenate the resulting liquids to further reduce the oxygen so that the final product consists mostly of hydrocarbons. To test this approach, three simulated catalytically upgraded pyrolysis liquids were prepared based on microscale tests of catalytic pyrolysis.. These liquids were then hydrotreated in a 100-mL batch autoclave at 250-360°C and 1500 psi initial hydrogen pressure in the presence of a carbon-supported palladium catalyst.

The product compositions showed that a general reduction in oxygen was achieved as well as hydrogenation. However, in order to achieve high levels of deoxygenation, the more severe condition (360°C) was needed to deoxygenate phenolic compounds. The main mechanism was hydrogenation followed by hydro-deoxygenation.

400025 Sustainable Energy Production from Renewable Resources through Anaerobic Digestion Process: Nigerian Experience

Abstract:

Nigeria is a country with abundant, diverse and un-exploited renewable energy resources that are yet to be used for improving the livelihood of the vast majority households. The production of biogas via anaerobic digestion of agricultural residues, municipal wastes and industrial wastes (water) would benefit Nigerian society by providing a clean fuel in the form of biogas from renewable feedstock and help end energy poverty. These renewable resources include water hyacinth, chicken litters, cattle dung and cattle rumen. The focus of this paper is on the conversion of these resources to sustainable energy by using anaerobic digesters. The anaerobic digester is made up of 10 liters plastic keg which was designed for the purpose of this batch pilot-scale laboratory research work. The height, width and length of the digester were 28cm, 16cm and 22.4cm respectively. The diameter of the inlet of the digester was 3.8 cm. The diameter and length of the outlet tap were 2.9cm and 11cm respectively. Two and a half kilogram of cattle dung was measured and seven liters of water was added to it to form slurry. The total volume was measured with a calibrated bucket. The composite weight was also measured with a weighing balance. The slurry was poured into the digester and was properly sealed. The same process and measure was used for chicken litter, cattle rumen and water hyacinth. After few days of anaerobic digestion, the digester was swollen, the gas inside was tested to confirm whether it was biogas; but it was majorly carbon-dioxide because it does not support combustion and thus, it was released. Biogas was later produced on the thirtieth day as the gas combusted. The gas was measured by using the water displacement method. The amount of gas produced was monitored for ten days. The result of the findings was recorded. Chicken litter, Water hyacinth and Cattle rumen were treated in the same way and biogas production started on the twenty-eight, twenty-sixth and twenty-third days respectively. Nigeria is a haven for Cattle wastes, Poultry dropping, Water hyacinth and Cattle rumen. The utilization of these resources will provide sustainable energy for Nigerians.

401012 Is CFD Ready for Being a Design Tool for Fixed-Bed Reactors?

Abstract:

Is CFD ready for being a design tool for fixed-bed reactors?

Gregor D. Wehinger, Thomas Eppinger, Matthias Kraume

Chemical and Process Engineering, Technische Universität Berlin, Fraunhoferstr. 33-36, 10587 Berlin, Germany

Almost 15 years ago Dixon and Nijesmeisland simulated for the first time a fixed bed with a small tube-to-particle-diameter ratio (*N*) that took the actual particle shape into account. Although the bed consisted of only some uniformly packed spheres, it was now possible to investigate the interstitial flow field without *a priori* assumptions for momentum transport. Consequently, the authors claimed "CFD as a design tool for fixed-bed reactors" [1]. After that groundbreaking paper, several scientists have extended these detailed CFD simulations toward non-spherical shaped particles, particles including thermal conductivity, as well as catalytic active particles. However, several aspects of an adequate modeling are still under investigation, i.e., among others turbulent flow, meshing criteria, radiation, and the coupling of heterogeneous catalysis on the particle surface with the surrounding flow field.

In this work we are looking critically at recent developments in the field of spatially resolved simulations of catalytic fixed-bed reactors. In addition, examples are given, how such simulations can be validated with appropriate detailed experimental techniques. Therefore, a capillary sampling technique is applied in a tubular fixed-bed reactor [2]. This method provides profiles of gas species concentrations, gas temperature and surface temperature. These experimental results are then compared with 3D numerical simulations taking explicitly into account the shape of the spherical particles. Inside the particles heat conduction is assumed. On the particle surface a detailed reaction mechanism is applied distinguishing between adsorption, surface reaction and desorption of chemical species. The dry reforming of methane (DRM) is taken as an example of heterogeneous catalytic processes. The procedure of the fixed-bed model follows that of Wehinger et al. [3]. This modeling approach allows the investigation of the interactions between chemical kinetics and transport of momentum, heat and mass.

Finally the applicability of CFD as a design tool is investigated. Therefore, three different kinds of particles of typical industrial dimensions are evaluated under common reaction conditions toward conversions and yields of the DRM. The particle shapes involve spheres, cylinders and 1-hole cylinders, see *Fig*. 1. With such detailed modeling transport phenomena inside fixed-beds can be quantified and finally, a substantial comparison between different packings is possible.

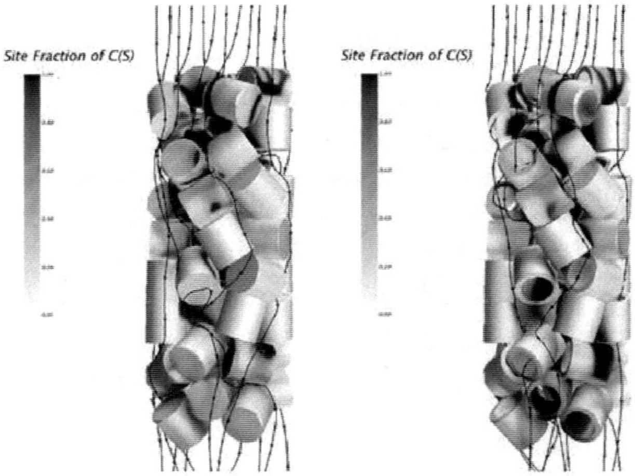

Fig. 1. Surface site fraction of adsorbed carbon C(S) on cylindrical particles (left) and 1-hole cylindrical partciles (right) for DRM.

However, the time required for a converged simulation can account for several days. Accordingly, such simulations are too time-consuming that they can replace lumped models, which deliver results in seconds, in the design stage. Nevertheless, detailed simulations of small *N* fixed-bed reactors offer unique insights into the complex processes of catalytic reacting flow. Moreover, dependencies on empirical equations can be minimized.

References

[1] Dixon, A. G. & Nijemeisland, M. *Ind. Eng. Chem. Res.*, **2001**, 40, 5246-5254

[2] Horn, R., Degenstein, N., Williams, K. & Schmidt, L. Catal. Lett., **2006**, *110*, 169-178

[3] Wehinger, G.D., Eppinger, T. & Kraume, M. Chem. Eng. Sci., **2015**, *122*, 197-209

MODELING THERMAL ENERGY STORAGE UTILIZING AIR AS HEAT TRANSFER FLUID AND ALUMINA AS STORAGE MATERIAL

Paper number 400964

Dr. Ryan Anderson, Montana State University, Bozeman, MT, USA
Liana Bates, Montana State University, Bozeman, MT, USA

ABSTRACT

Thermal energy storage is a critical component of concentrated solar power plants to increase the capacity factor and drive down the cost of electricity. One storage option is a packed bed of a solid storage material with a gaseous heat transfer fluid. Alumina is considered a strong candidate as the storage material in the packed bed due to a high heat capacity and thermal stability, making it well suited for high temperature thermal energy storage. A simplified, one-equation model is presented with air as heat transfer fluid flowing through a packed bed of α-alumina beads. The thermal properties are averaged in the one-equation model, and all of these properties are calculated as a function of temperature. In addition to the thermal model, the pressure drop is also calculated. The modeling approach is compared to experimental data at multiple flow rates during storage and recovery. For this air and alumina system, the one-equation approach works well in describing the thermal front as it moves through the packed bed. With this validated approach, a designer could quickly study the key parameters (e.g. bed diameter, bead size, materials) associated with the storage for a user's specific conditions (e.g. maximum temperature, flow rates).

INTRODUCTION

Concentrating solar power (CSP) plants are one sustainable energy option that can generate electricity from solar energy. In contrast to photovoltaic systems, these systems concentrate solar energy by reflecting sunlight to a receiver. At the receiver, a 'thermal energy carrier' is heated, which is then utilized to generate power [1]. Thermal energy storage (TES) can be incorporated to offset variability [1,2] and reduce costs [3,4]. Storage and recovery must be done efficiently to achieve high capacity factors/low LCOE. Several thermal energy storage designs exist [5]. TES technologies must meet several requirements [6]: high energy density, good heat transfer between the heat transfer fluid (HTF) and solid storage media, stability of the storage material, low thermal losses, low cost, and reversibility through charging and discharging cycles. The present study explores sensible thermal energy storage where the solid storage design is a packed bed [7,8]. To charge the bed, the heat transfer fluid heated by solar radiation flows through the packed bed to heat the storage material. To recover the stored energy from the bed (discharge), the flow direction is reversed and low temperature HTF enters the bed. The exiting HTF is at the high temperature and can then be used in a power cycle. The storage step is shown schematically in Figure 1a, while the experimental schematic for the data included in this summary is in Figure 1b [9,10].

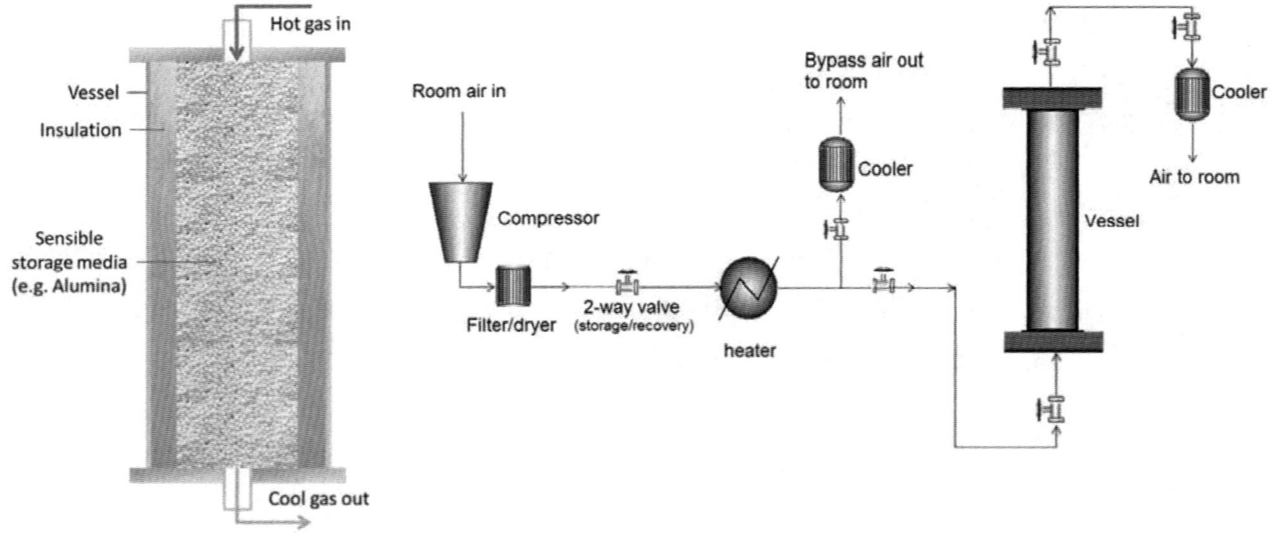

Figure 1 a) Schematic of packed bed thermal energy storage where the packed bed is contained within a layer of internal insulation and a pressure vessel. The hot gas enters, deposits heat to the bed, and exits at T_{cool}. b) Schematic of the experimental setup presented in [9], printed in [10], showing the direction of storage where room air is sent from a compressor through a filter/dryer, into a heater to simulate CSP, through a bypass valve to set the flow rate, and into the TES vessel.

The primary focus of this work is with α-alumina as the solid material. Air is the main focus as heat transfer fluid since experimental data exists on this system. Model results from software packages COMSOL Multiphysics and Star-CCM+ are compared to existing experimental data. The model approach in all cases uses a one-equation energy model to predict the thermal behavior in the packed bed. The pressure drop is also calculated, though these results are not presented here. To highlight the general applicability of the modeling approach, recent work with supercritical CO_2 as the heat transfer fluid with alumina as the storage material is presented. Future work with an air-rock TES system in collaboration with the University of Stellenbosch is also presented briefly.

NOMENCLATURE

ρ	density (kg/m^3)	t	time (s)
c_p	heat capacity (J/kg K)	$\bar{u}$	superficial gas velocity (m/s)
T	temperature (K)	Q_{loss}	heat loss (W)
ε	porosity	k	thermal conductivity (W/m K)

RESULTS AND DISCUSSION

The one-equation model is presented first along with a brief explanation of the experimental facility used to collect the data. Two key modeling results are then shown: 'short-storage' results solved with Star-CCM+ and 'long-storage' results with COMSOL. Then, on-going work with supercritical CO_2 as heat transfer fluid is shown, followed by initial work with rocks as the storage media.

Model development

The data collected (and utilized here again) was initially modeled with a two-equation approach, which considered energy balances for the fluid and solid in the packed bed [9,11]. One drawback of that approach is that it relied on a heat transfer coefficient between the fluid and solid in the packed bed, and determining the heat transfer coefficient relied on inaccurate empirical correlations. In this presentation, a one-equation approach to the energy balance is used, which uses averaged properties between the two phases instead of using a heat transfer coefficient. For the air-alumina system, thermal equilibrium can be reasonably assumed, making this approach valid [9,12].

In the packed bed domain, the energy balance equation is:

$$(\rho c_p)_{eq}\frac{\partial T}{\partial t} + \rho c_p \mathbf{u}\cdot\nabla T = \nabla\cdot(k_{eq}\nabla T) + Q_{loss} \quad (1)$$

where the equivalent thermal properties (subscript *eq*) are averaged based on the solid and fluid fractions:

$$(\rho c_P)_{eq} = (1-\varepsilon)\rho_s c_{p,s} + \varepsilon\rho_f c_{p,f} \quad (2)$$

$$k_{eq} = (1-\varepsilon)k_s + \varepsilon k_f \quad (3)$$

The energy equations for the solid domains (insulation and storage vessel) are:

$$(\rho c_p)_i \frac{\partial T}{\partial t} + = \nabla\cdot(k_i \nabla T) \quad (4)$$

The solver also includes mass conservation (continuity equation) and momentum conservation to predict the pressure drop, but those equations are not included here as the following results focus on the thermal behavior of the packed beds.

Initially, the domains are set to a constant temperature. A natural convection term is used as a boundary condition on the external vessel wall to simulate heat loss. Additional boundary conditions include a fixed pressure at the exit and an inlet mass flux (kg/m^{2}s) or mass flow rate (kg/s).

α-alumina is a suitable storage material [13] due to advantageous physical properties, mechanical stability, and thermal stability. Note that accurate simulations must include the effect of temperature on the thermophysical properties. The relationships for each property as a function of temperature from [9, 10] are used in this work for air and alumina.

The equations were solved with commercial software packages COMSOL and Star-CCM+. The student completed the COMSOL portion as part of an independent study for course credit. Additional studies, with a specific focus on research instead of teaching, were completed in Star-CCM+.

Results for air as heat transfer fluid with alumina as storage material

Data was collected at an experimental facility at CCNY [9,11] with a packed bed that is 2.25 inches in diameter and 10 ft tall.

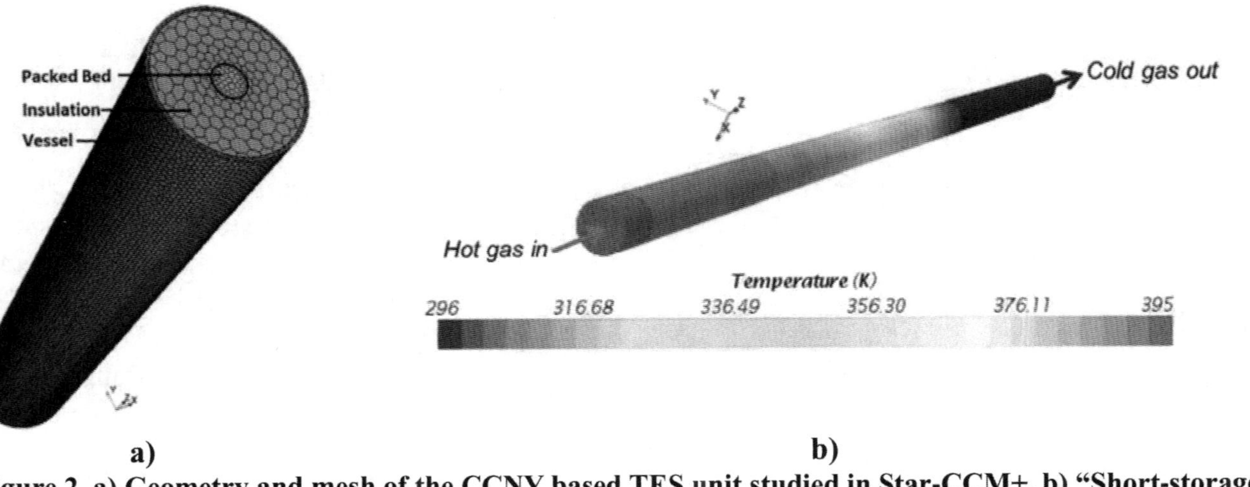

Figure 2. a) Geometry and mesh of the CCNY based TES unit studied in Star-CCM+. b) "Short-storage" simulation results highlighting the axial and radial temperature profile of the bed solved by Star-CCM+.

The geometry of that system, as input into Star-CCM+ is shown in Figure 2a for illustrative purposes. Highlighted are the packed bed, interior insulation, and steel wall. Figure 2b highlights a typical temperature result where only the packed bed temperature is isolated. Note, this allows one to view both the axial and radial temperature profiles, which is important in determining the overall storage and recovery efficiencies.

Experimental data and modeling results for 6 mm alumina beads and air flow at 10 SCFM are shown in Figure 3a. Note for this data a dimensionless temperature is used with $\Theta = 1$ corresponding to the hot storage temperature. An explanation of all parameters are listed in ref. [10], as this data set was used for validation in that study. These 'long-storage' results are from two hours of data collection. COMSOL solved the equations listed above, and overall the agreement is quite good. The model was also compared to data for 20 minutes of storage, and these results are in Figure 3b. Star-CCM+ was used to obtain these results. Again, the data corresponds to 6 mm alumina beads, but the flow rate here is 20 SCFM. Reasonable agreement between the data and the model is noted for these initial efforts, though the model's thermal front appears sharper than the experimental data.

Some properties were assumed, such as the insulation properties and the heat transfer coefficient for natural convection at the vessel boundary. Heat losses to the peripheral piping and vessel flanges during start-up are accounted for via a ramp in the temperature boundary condition at the bed inlet. This ramp was estimated via the experimental data in both cases.

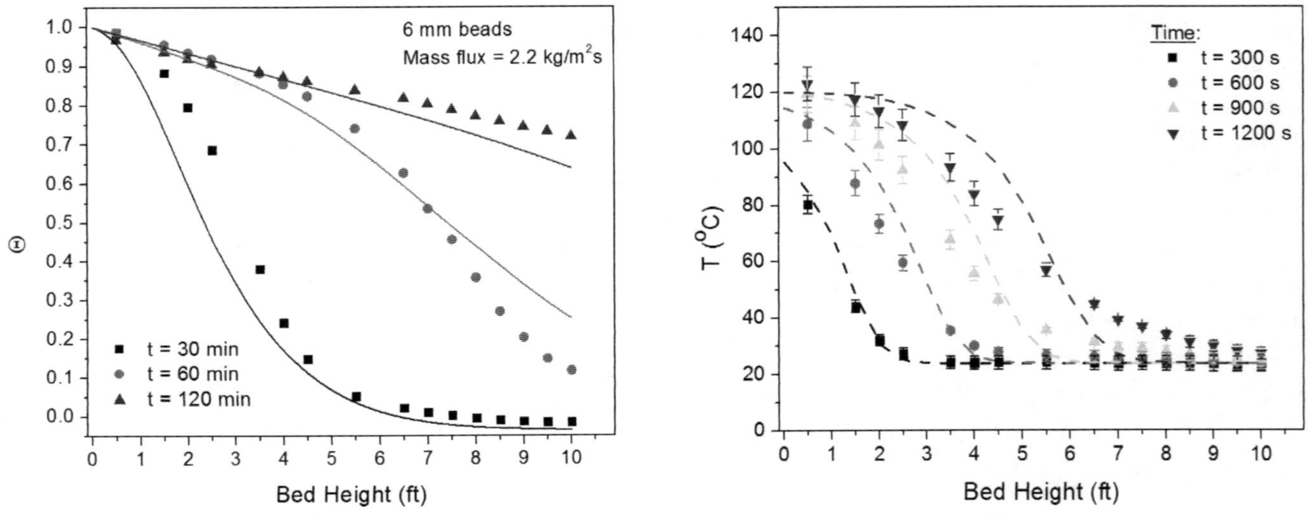

a) b)

Figure 3. a) "Long-storage" simulation results for the model (solid line) vs. experimenetal data over two hours [10] solved by COMSOL. b) "Short-storage" simulation results for the model (dashed line) vs. experimenetal data over 1200s (20 minutes) solved by Star-CCM+.

Applications of alumina with supercritical carbon dioxide as heat transfer fluid

The modeling approach here is not limited to air as heat transfer fluid and alumina as storage material. Recent work explored supercritical CO_2 (sCO_2) as heat transfer fluid for use in a recuperated sCO_2 Brayton cycle [10]. The conditions were based on goals from the Department of Energy [14]. This led to a design for a packed bed of alumina spheres with an inlet temperature of 720°C at approximately 275 bar. Similar to the previous study, a cylindrical containment vessel was used with alumina spheres. The packed bed was simulated as 3.5 m in diameter and 9 m tall. The properties of sCO_2 varied as a function of temperature [15]. The results considered storage over six hours.

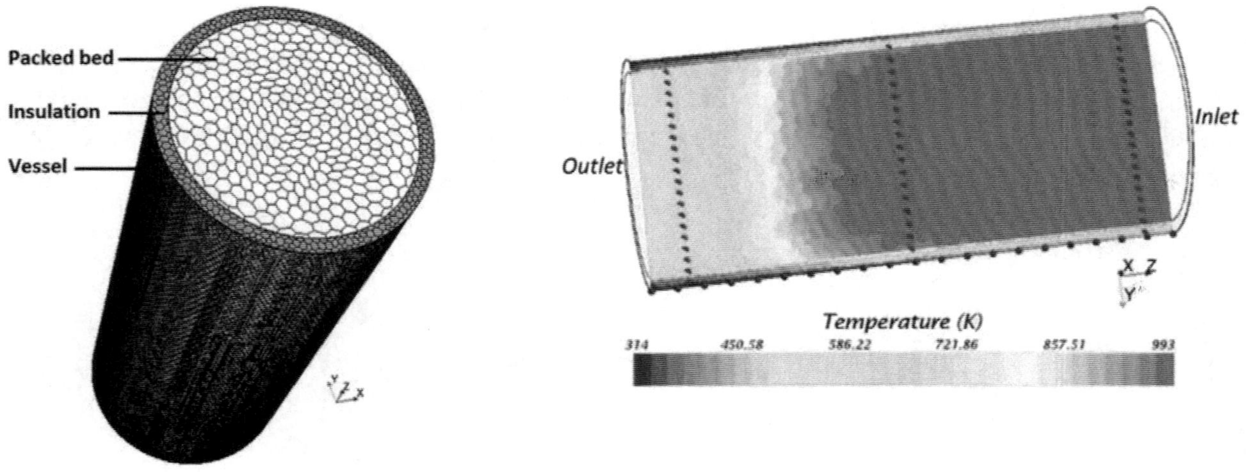

Figure 4. a) Schematic of geometry and mesh solved in Star-CCM+ for a d = 3.5 m, h = 9 m vessel. b) Axial and radial temperature profile during storage for a center cut through the packed bed.

Figure 5a shows the temperature profile in the packed bed at the center of the vessel, solved by both COMSOL and Star-CCM+. In general, the model predictions are similar. For practical use, a reasonably sharp thermal front is noted, indicating high energetic and exergetic efficiencies may be possible in this configuration.

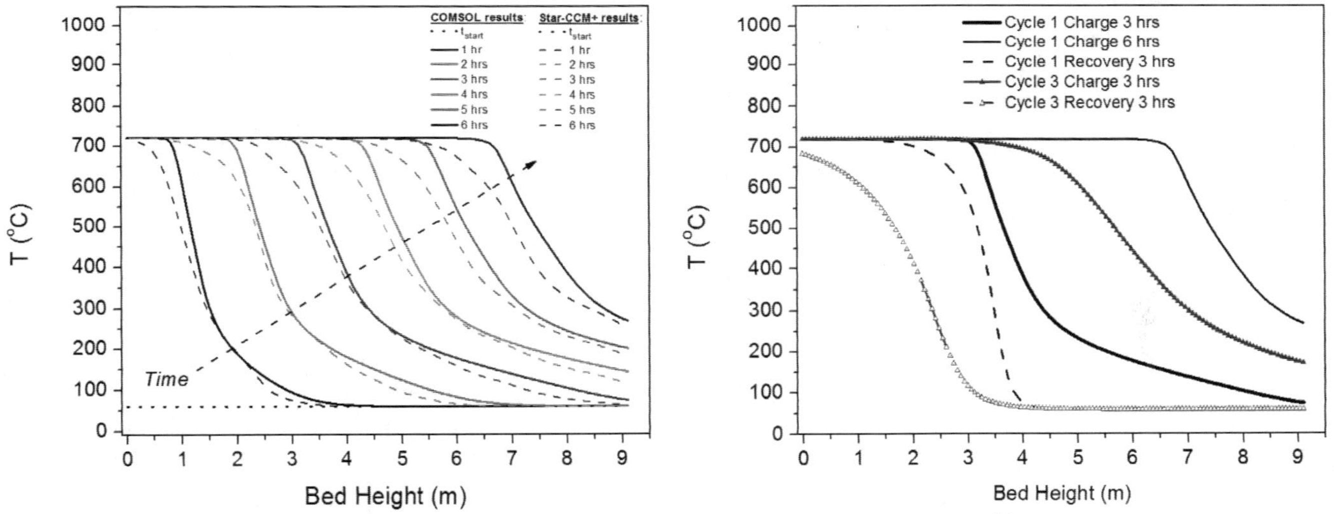

a) b)

Figure 5. a) Thermal front moving through the bed at the centerline of the alumina TES unit during a six-hour storage period. b) Temperature profiles during three cycles starting from an initially cold temperature. Both results are presented and thoroughly discussed in [10].

Figure 5b shows the temperature profiles for the first and third cycles, where each cycle is 6 hours of charging (solid lines) and 3 hours of recovery (dashed lines). At the third cycle, less of the bed is available for additional storage and the thermal front is not as sharp, which means the efficiency of the unit decreases with cycling. Though this behavior is expected, a more thorough design and well-understood control scheme would be necessary to use this system commercially.

Applications of rocks as storage material with air as heat transfer fluid

Not only is the approach not limited to the heat transfer fluid, the storage material can also be varied. Initial modeling has started to compare with results from experimental data collected at the University of Stellenbosch. A schematic of the modeling domain is shown in Figure 6a, which matches the experimental facility.

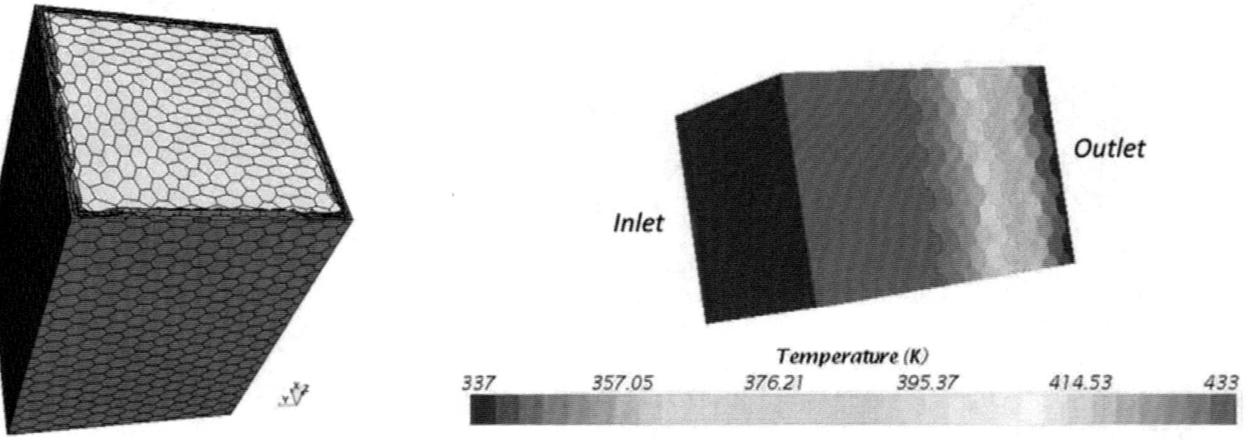

a) b)

Figure 6. a) Geometry and mesh created in Star-CCM+ for the air-rock system. b) Initial temperature results showing the 3D temperature profile of the rectangular packed bed.

It is hoped that the 3D analysis will complement their previous modeling efforts [16]. Dolerite is the packed bed solid material with an average equivalent diameter of approximately 39 mm. The heat capacity changes as a function of temperature. This vessel is rectangular with dimensions of 1x1x1.5 m. Thin layers of insulation are placed on the inside and outside of the storage vessel. Future work will compare the modeling results to the experimental data, which includes not only the axial temperature profile in the packed bed but also the pressure drop data.

CONCLUSIONS

Thermal energy storage is a technology receiving increased attention in applications like concentrated solar power because it can extend the plant's capacity factor, driving down the cost of electricity. One system for energy storage uses sensible heat storage with a solid in a packed bed arrangement. Hot fluid passes through the bed, depositing heat onto the solid storage media. To recover the energy, cold fluid enters the bed, heat is transferred to the fluid, and hot fluid exits the bed. The hot fluid can then be utilized in a power cycle. To design these systems for high efficiency, several modeling options exist. For the packed beds described here, a one-equation model is appropriate, which solves a single energy equation with averaged properties from the fluid and solid. A system of air as HTF and alumina as storage material is first presented, as experimental data exists from a facility at CCNY. Both COMSOL and Star-CCM+ are used to compare model results to this data. Overall, agreement is reasonable. The approach is broadly applicable, and two other systems are presented. One system uses alumina as storage media and sCO2 as HTF at high temperature. This system exhibits a sharp thermal front, but more research is needed to design an optimized system. Another system uses air as HTF and local rocks as the solid storage media, and this study is the focus of upcoming work.

ACKNOWLEDGMENTS

We would like to thank Dr. Erick Johnson (Montana State University) for his assistance in Star-CCM+ and Dr. Pablo Bueno (Southwest Research Institute) for his expertise in the supercritical CO_2

Brayton cycle. Additional thanks are also given to the Undergraduate Scholars Program at Montana State University.

REFERENCES

[1] Zhang, H. L., Baeyens, J., Degrève, J., & Cacères, G. (2013). Concentrated solar power plants: Review and design methodology. Renewable and Sustainable Energy Reviews, 22, 466-481.

[2] Li P. Energy storage is the core of renewable energy technologies. IEEE Nanotechnol Mag. 2008;4:3-18.

[3] International Renewable Energy Agency, Renewable Energy Technologies: Cost Analysis Series, Volume 1: Power Sector Issue 2/5, Concentrating Solar Power, June 2012.

[4] US Department of Energy, SunShot vision study. Chap. 5. Feb. 2012, p. 115.

[5] Li P, Lew JV, Chan C, Karaki W, Stephens J, O'Brien JE. Similarity and generalized analysis of efficiencies of thermal energy storage systems. Renewable Energy 2012;29:388-402.

[6] Kuravi, S., Trahan, J., Goswami, D. Y., Rahman, M. M., & Stefanakos, E. K. (2013). Thermal energy storage technologies and systems for concentrating solar power plants. Progress in Energy and Combustion Science, 39(4), 285-319.

[7] Coutier JP, Faber EA. Two applications of a numerical approach of heat transfer process within rock beds. Sol. Energy 1982;29:451-62.

[8] Zanganeh G, Pedretti A, Zavattoni S, Barbato M, Steinfeld A. Packed-bed thermal storage for concentrated solar power – Pilot-scale demonstration and industrial-scale design. Sol. Energy 2012;86:3084-98.

[9] Anderson, R., Shiri, S., Bindra, H., & Morris, J. F. (2014). Experimental results and modeling of energy storage and recovery in a packed bed of alumina particles. Applied Energy, 119, 521-529.

[10] Bueno, P.C, Anderson, R., Bates, L., Bindra., H. Thermal energy storage for the supercritical CO_2 Brayton cycle, Proceedings of ASME Turbo Expo 2015: Turbine Technical Conference and Exposition, GT2015-44054.

[11] Samira Shiri, MS Thesis, City College of New York, A model of transient heat transfer in a packed bed of alumina particles

[12] Nield, D. A., & Bejan, A. (2006). Convection in Porous Media. springer.

[13] Bindra, H., Bueno, P., Optimum process design of packed bed type thermal storage systems and other applications, WO patent 2013015834.

[14] Department of Energy, Office of Energy Efficiency and Renewable Energy (EERE), Concentrating Solar Power: Advanced Projects Offering Low LCOE Opportunities, Funding Opportunity Announcement (FOA) Number: DE-FOA-0001186, 30 September 2014.

[15] E.W. Lemmon, M.O. McLinden and D.G. Friend, "Thermophysical Properties of Fluid Systems" in NIST Chemistry WebBook, NIST Standard Reference Database Number 69, Eds. P.J. Linstrom and W.G. Mallard, National Institute of Standards and Technology, Gaithersburg MD, 20899, http://webbook.nist.gov, (retrieved December 19, 2014).

[16] Allen, K. G. (2010). Performance characteristics of packed bed thermal energy storage for solar thermal power plants (Doctoral dissertation, Stellenbosch: University of Stellenbosch).

401324 Computational Modelling of Dimethyl Ether (DME) Steam Reforming in a Dual Fluidized Bed System

Abstract:

DME has attracted attentions as a potential source of hydrogen for fuel cells. This study presents a theoretical investigation of the catalytic steam reforming of dimethyl ether (DME) for the production of hydrogen in the riser section of a dual fluidized bed (DFB) system. The DFB has the advantages of sustainable operation while regenerating the catalysts within the closed loop system. A two-phase (Eulerian-Eulerian) model was solved in CFD software (ANSYS FLUENT) to simulate the reactor flow hydrodynamics, heat transfer and reactions in three dimensional coordinates. A sub-model for the reactions and kinetics was developed and incorporated in the model as user defined function (UDF). The kinetic parameters for the DME steam reforming and the associated methanol reforming and water gas shift reactions were obtained from the experimental data reported by Feng et al., (2009) and Yan et al., (2014) with bi-functional catalyst $CuO/ZnO/Al_2O_3$ -ZSM-5. The effect of steam and catalysts to DME ratio, reactor temperature and space velocity on the DME conversion and product gas composition were investigated. The results have shown strong dependence of the DME conversion on the reactor operating temperature with the maximum hydrogen yield and 100% DME conversion obtained at the temperature >300 °C. The parametric analysis have shown good match between the model predictions and the reported experimental data, thus demonstrating the great potential of the proposed model for optimization and future development of the proposed process to industrial scale applications.

Feng, D., Wang, Y., Wang, D., & Wang, J. (2009). Steam reforming of dimethyl ether over CuO–ZnO–Al2O3–ZrO2+ZSM-5: A kinetic study. *Chemical Engineering Journal*, *146*(3), 477–485. doi:10.1016/j.cej.2008.11.005

Yan, C.-F., Ye, W., Guo, C.-Q., Huang, S.-L., Li, W.-B., & Luo, W.-M. (2014). Numerical simulation and experimental study of hydrogen production from dimethyl ether steam reforming in a micro-reactor. *International Journal of Hydrogen Energy*, 4–11. doi:10.1016/j.ijhydene.2014.02.133

400687 Hydrolysis of Biomass Using a-Hydroxysulfonic Acids

Abstract:

Due to current practices in the petroleum industry, fossil fuel availability is permanently decreasing while their use significantly contributes to the accumulation of greenhouse gasses. The sustainability and feasibility for the conversion of biomass to hydrocarbon fuels has therefore become an essential research focus. While fuels are burned for energy and release CO_2 into the atmosphere, replanting previously harvested biomass guarantees that CO_2 will be recycled back through the growth of new foliage.

Biomass can be converted into valuable chemicals through methods such as acid catalyzed hydrolysis and dehydration. The cellulose in biomass is hydrolyzed into a few useful chemicals, including levulinic and formic acid, which can be further upgraded to liquid fuels through the relatively simple thermal deoxygenation (TDO) method. Currently, the preferred method of acid catalyzed hydrolysis of biomass is performed with sulfuric acid. It is necessary to recover the sulfuric acid for both economical and environmental reasons. The recovery and recycling of the sulfuric acid has proven to be a significant hurdle with many challenges affecting the scale-up and commercialization of the biomass hydrolysis processes.

Hydrolysis of biomass using a class of acids called α-hydroxysulfonic acids presents a novel approach to simplifying acid recovery and recyclability. Until now, little research has gone into hydroxysulfonate formation outside of the chemical pulping field. We have shown that specific α-hydroxysulfonic acids can be intentionally synthesized to effectively hydrolyze biomass to levulinic and formic acid. The acid and its constituents exist in equilibrium with each other. Therefore, the acid can be easily reverted completely to its non-acidic constituents by employing the proper conditions. The position of equilibrium is subject to the nature of the constituents used to synthesize the acid. The ability to tune the α-hydroxysulfonic acids in response to desired acid strength and decomposition method becomes a viable approach to process design. Complications in downstream processing can be minimized as separations, neutralization, and purification steps can be simplified or avoided. This study presents the results of an investigation into the use of several different α-hydroxysulfonic acids for conversion of cellulose to levulinic and formic acids.

399354 Effect of Water Content on Soybean Oil Transesterification: Design of Experiment

Abstract:

Effect of water content on soybean oil transesterification: design of experiment

Min-Hong Chen, Jia-Ming Chern*

Department of Chemical Engineering

Tatung University

Taipei, Taiwan

Abstract

Biodiesel, produced from transesterification of vegetable or animal oils and fats with alcohol (methanol or ethanol), is one of promising substitute fuels in the future. The predominant method for commercial biodiesel production is the base-catalyzed transesterification process using a low-cost catalyst such as sodium hydroxide or potassium hydroxide. In the base-catalyzed process, even the original oil contains no moisture; upon the contact of methanol with sodium hydroxide water is inevitably formed and causes troubles in the transesterification process. This study added molecular sieve during transesterification to remove water content and thus increased the biodiesel yield. The full factorial experimental design approach was employed to find the main and interactional effects first, after that the simple method was used to find the optimal condition for biodiesel yield.

401664 Measurement of Internal Temperature of a Li-Ion Cell at Large Discharge Rates

Abstract:

Lithium-ion cells [1-2] have been investigated for a variety of energy storage and conversion processes in several applications including military systems, consumer electronics and electric vehicles [3-5]. The process of conversion of electrical energy to chemical energy and vice versa inherently produces heat [6-9], which results in increased cell temperature [10-12]. The operation of a Lithium-ion cell is based on several highly coupled phenomena involving multiple physical processes. For example, exothermic electrochemical reactions produce heat [6, 9, 13], which causes temperature rise. Since temperature directly affects the rates of electrochemical reactions and electrical impedances [6], therefore, the thermal performance of the cell in turn affects electrochemical and electrical performance. Cell performance, cycle life, and system safety all depend on temperature distribution in the cell, which, in turn, depends on heat generation rate within the cell and on heat removal rate at the cell surface.

This paper presents a method for measurement of internal temperature of a Li-ion cell. The method is based on measurement of the outside temperature and extrapolation to the internal temperature based on the linear nature of the governing energy equations, and using a priori measurements of the internal and external temperatures on one cell in which a small hole has been drilled along the axis for internal temperature measurement. Measurements are found to be in good agreement with the theoretical model presented in the paper. Results illustrate a powerful way for indirect measurement of cell temperature during operation, and may be helpful in the design of thermal management strategies for high-rate electrochemical energy conversion.

UNIVERSITY of DAYTON

Characterization of Phthalocyanine-Based Materials for Lithium-Ion Batteries

AIChE 2015 Spring Meeting
Austin, TX
April 26-30, 2015

Sarwan S. Sandhu*
Department of Chemical & Materials Engineering
University of Dayton, Dayton, OH 45469, USA

Joseph P. Fellner
Air Force Research Laboratory
Wright-Patterson AFB, OH 45433-7251, USA

1

Overview

- Introduction
- Theoretical
- Experimental Cells
- Discussion of Results
- Conclusions
- Acknowledgements

2

Introduction

Rechargeable lithium-ion battery

- For the storage of electric energy produced in (excess of its demand) by any electric energy producing system, a rechargeable battery, such as a lithium-ion battery is required.

Schematic of a Lithium-Ion Cell

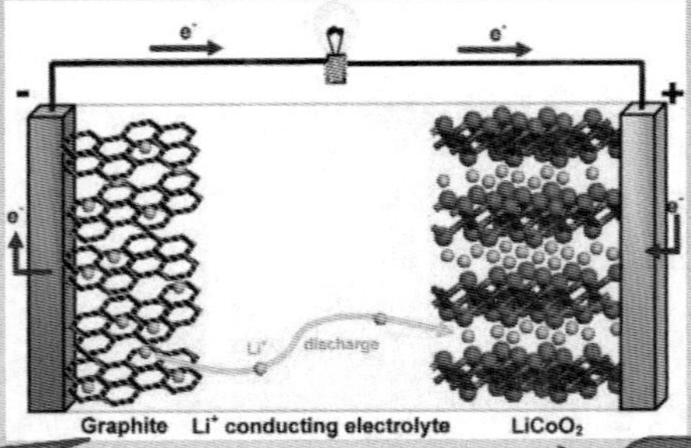

Graphite Anode Active Material:
Theoretical Capacity: 372 mAh/g
Actual Capacity: ~350 mAh/g

Lithium Cobalt Oxide Cathode Active Material:
Theoretical Capacity: 274 mAh/g
Actual Capacity: ~140 mAh/g

Introduction (continued)

Lithium/iron phthalocyanine battery

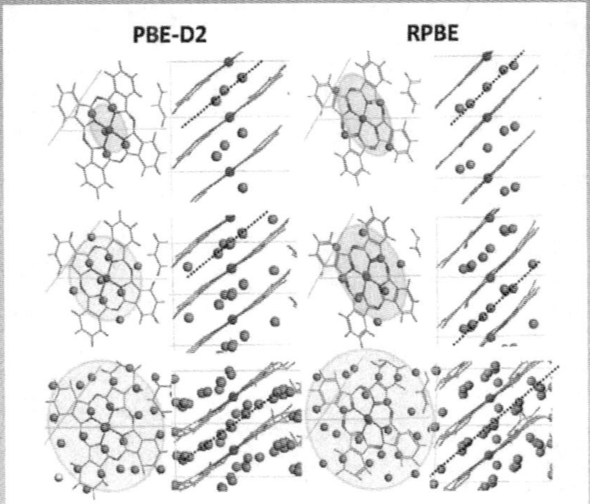

Fig. 2 Position of Li atoms intercalated in FePc calculated with the indicated exchange correlation functional (Fe in red, N in blue, C in gray and Li in purple). In the top view the red circle represents the area within which Li atoms are confined, in the side view the dotted line is located parallel to the middle of the planes formed by FePc molecules.

Source: G. Ramos-Sanchez, A. Callejas-Tovar, L.G. Scanlon, and P.B. Balbuena, *Phys. Chem. Chem. Phys.*, 2014, 16, 743

4

Theoretical

- Technical work presented here is on a metal phthalocyanine (for example, iron phthalocyanine) as the cathode charge storage material
  - Theoretical charge storage capacity of FePc:
    - Up to 37 lithium atoms per FePc[1] molecule or 1745 mAh/gram of FePc
  - Practical reversible charge storage capacity of FePc:
    - Our current engineering judgment: about 20 lithium atoms per FePc molecule or 943 mAh/gram of FePc

1. G. Ramos-Sanchez, A. Callejas-Tovar, L.G. Scanlon, and P.B. Balbuena, *Phys. Chem. Chem. Phys.*, 2014, 16, 743

5

Theoretical (continued)

Thermodynamic functions to characterize the FePc material

- The overall process for lithium insertion into the cathode active material:

$$xLi_{(s)} + FePc_{(s)} \rightarrow Li_xFePc_{(s)}$$

- Thermodynamic functions:

$$\Delta G = \left[G_{Li_xFePc_{(s)}} - (G_{FePc_{(s)}} + xG_{Li_{(s)}}) \right]$$

$$= -F\int_0^x E_{rev}dx$$

6

Theoretical (continued)

Thermodynamic functions to characterize the FePc material

- Thermodynamic functions (cont'd):

$$\Delta S = \left[S_{Li_x FePc_{(s)}} - (S_{FePc_{(s)}} + x S_{Li_{(s)}}) \right]$$

$$= F \int_0^x \left(\frac{\partial E_{rev}}{\partial T} \right)_x dx$$

$$\Delta H = \left[H_{Li_x FePc_{(s)}} - (H_{FePc_{(s)}} + x H_{Li_{(s)}}) \right]$$

$$= F \left[-\int_0^x E_{rev} dx + T \int_0^x \left(\frac{\partial E_{rev}}{\partial T} \right)_x dx \right]$$

7

Theoretical (continued)

- Proposed mathematical expression for the determination of lithium ion diffusion coefficient in the solid FePc

$$D(x,T) = D_0 x^n \exp\left(\frac{-E_d(x)}{RT} \right)$$

8

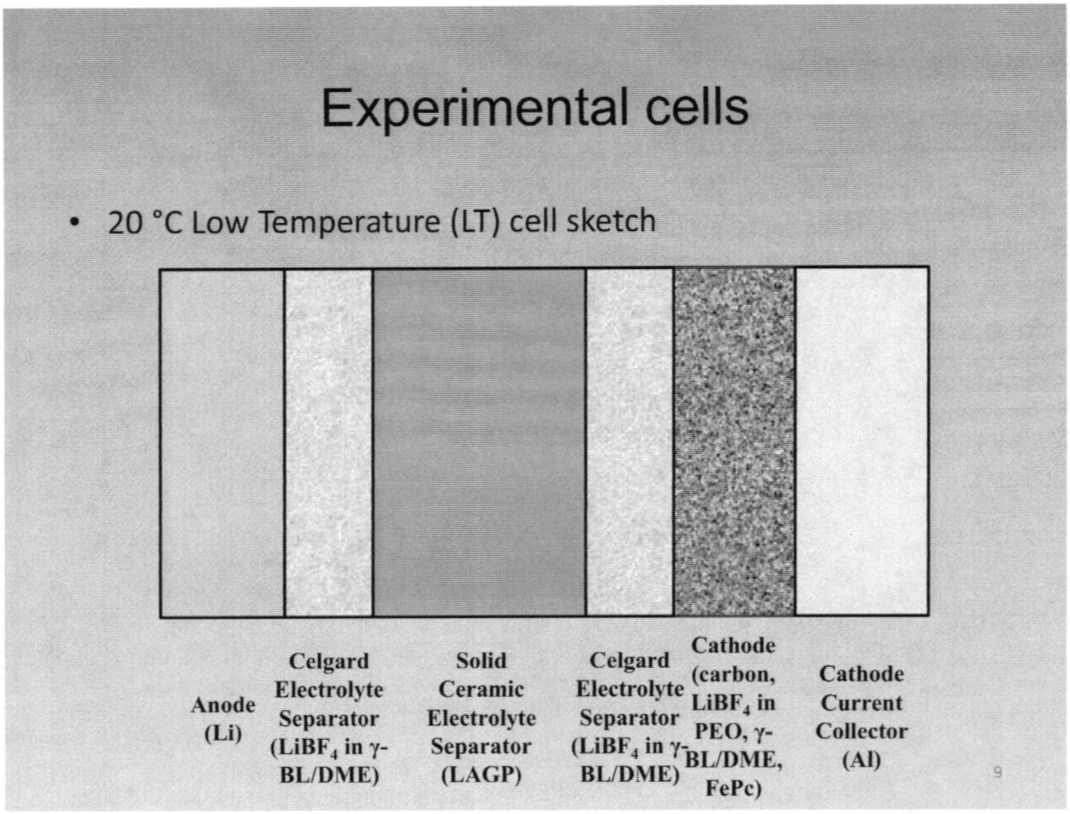
Experimental cells
• 20 °C Low Temperature (LT) cell sketch
Anode (Li)
Celgard Electrolyte Separator ($LiBF_4$ in γ-BL/DME)
Solid Ceramic Electrolyte Separator (LAGP)
Celgard Electrolyte Separator ($LiBF_4$ in γ-BL/DME)
Cathode (carbon, $LiBF_4$ in PEO, γ-BL/DME, FePc)
Cathode Current Collector (Al)
9

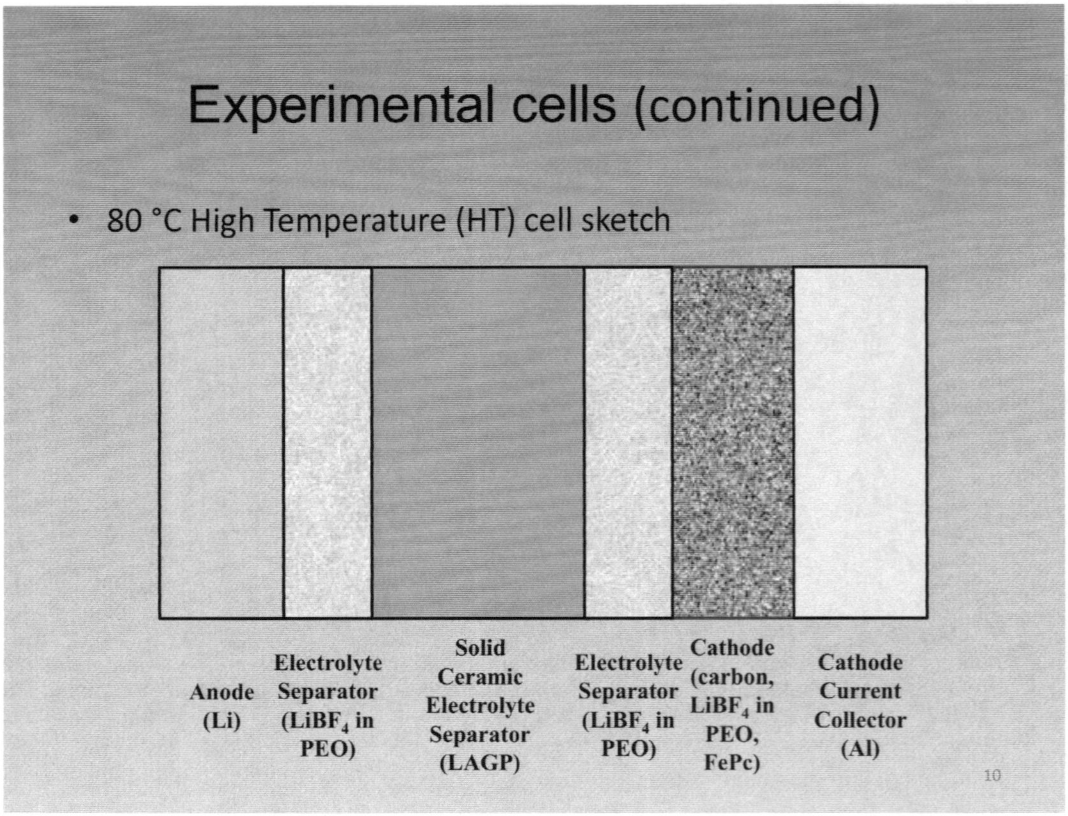
Experimental cells (continued)
• 80 °C High Temperature (HT) cell sketch
Anode (Li)
Electrolyte Separator ($LiBF_4$ in PEO)
Solid Ceramic Electrolyte Separator (LAGP)
Electrolyte Separator ($LiBF_4$ in PEO)
Cathode (carbon, $LiBF_4$ in PEO, FePc)
Cathode Current Collector (Al)
10

Experimental cells (continued)

Alternative 20 °C Low Temperature (LT) coin cell

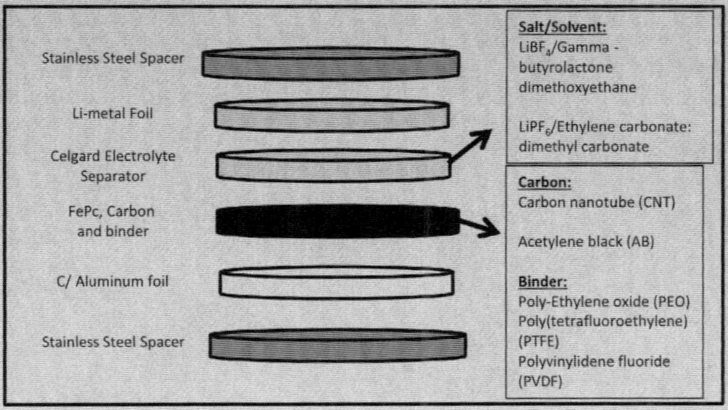

11

Results/Discussion

Two LT FePc coin cells discharge voltage as a function of delivered capacity on the first discharge

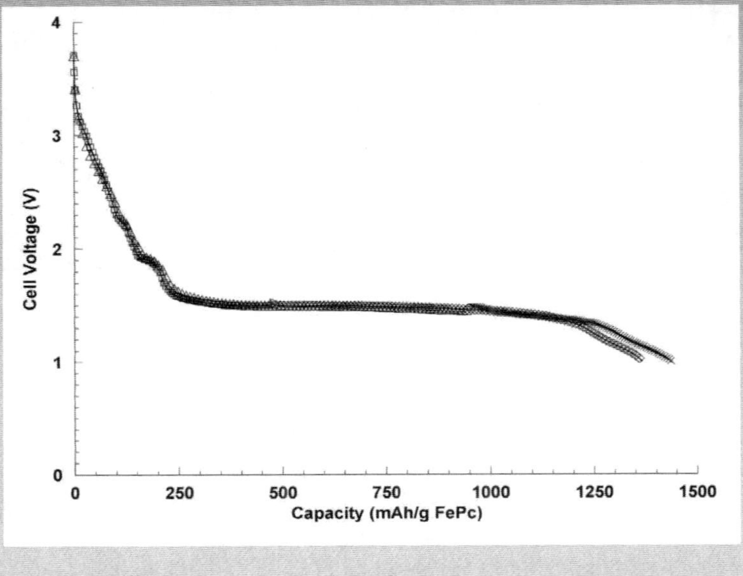

12

Results/Discussion (continued)

Two HT FePc coin cells discharge voltage as a function of delivered capacity on the first discharge

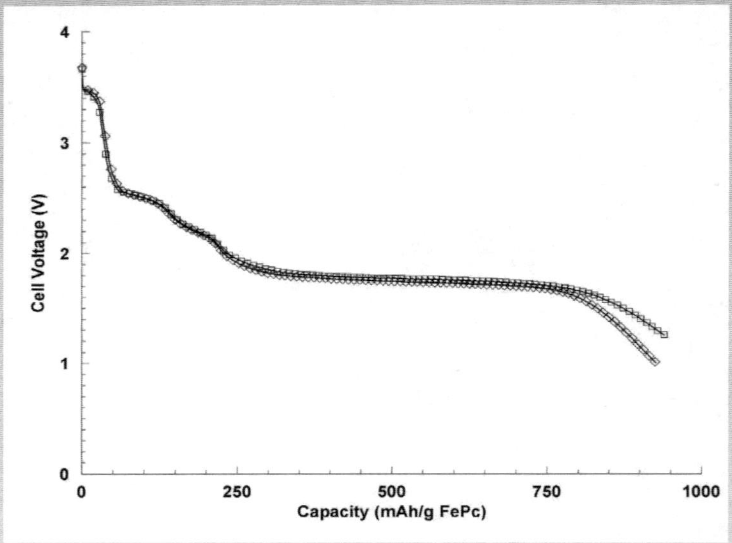

13

Results/Discussion (continued)

- Calculation of (-ΔG) as a function of x at 80 °C:

$$(-\Delta G) = F\left[\int_0^{x} E_{rev}dx\right] =$$

$$F\left[\int_0^{x_1} E_{rev}dx + \int_{x_1}^{x_2} E_{rev}dx + \int_{x_2}^{x_3} E_{rev}dx + \int_{x_3}^{x_4} E_{rev}dx\right]$$

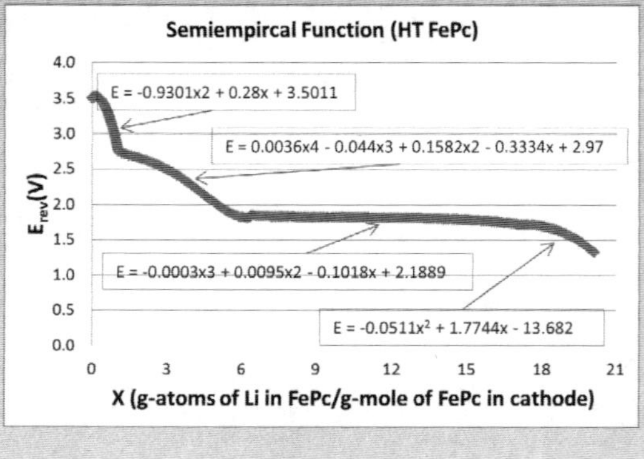

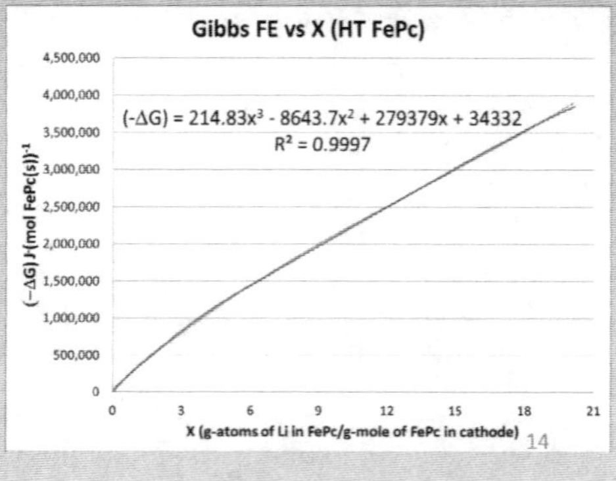

14

Results/Discussion (continued)

Cell impedance as a function of temperature

100% SOC, Li/$LiBF_4$ in PEO/LAGP/$LiBF_4$ in PEO/FePc coin cell

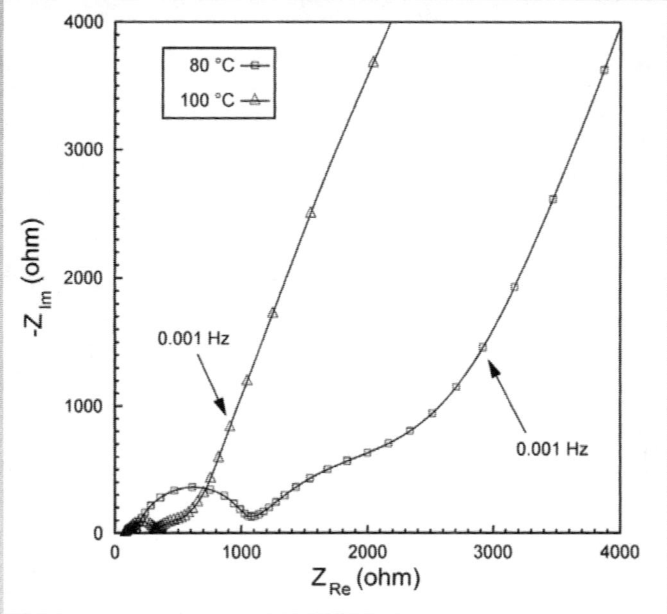

15

Results/Discussion (continued)

- Diffusion coefficient data
  - 80 °C, D = 7.74 x 10^{-12} cm^2/sec
  - 100 °C, D = 3.08 x 10^{-11} cm^2/sec
  - diffusion activation energy = 75.7 kJ/mol-K

16

Results/Discussion (continued)

Cycling of LT FePc coin cells

- Preliminary results show that the LT FePc cells can cycle at 20 equivalents or 943 mAh/g FePc at a C/10 discharge rate
  - the specific capacities of $LiMn_2O_4$, $LiCoO_2$ and $LiNiO_2$ are 126, 149 and 180 mAh/g, respectively
  - competitive with Li/S and Li/O_2 cells
  - work being done to increase reversible capacity and charge/discharge rate

17

Conclusions

- Experimental data on the cell voltage versus capacity for a high temperature FePc cell at 80 °C and a low temperature FePc cell at 20 °C acquired at approximately 40 µA (~C/100 rate) were presented.
- Empirical functions to simulate the cell voltage, E, versus x behavior during the various cell discharge stages were developed for a high temperature cell using the experimental data.

18

Conclusions (continued)

- In turn, such developed functions were used to develop an empirical equation for the determination of the Gibbs free energy of the reaction of lithium insertion into FePc as a function of x for a HT FePc operated at 80 °C.
- The lithium/FePc cell impedance experimental data in the form of Nyquist plots were employed to evaluate the lithium diffusion coefficients in the cell cathode active material, FePc(s).
- The experimentally determined diffusion coefficients for the lithium ion penetration into the cathode active material are: 7.7×10^{-12} and 30.8×10^{-12} cm2 s^{-1} at 80 and 100 °C, respectively, with the diffusion activation energy, $E_d = 75.7$ kJ mol^{-1}.

19

Conclusions (continued)

- Preliminary results show that the LT FePc cells can reversibly cycle at 20 equivalents or 943 mAh/g FePc at a C/10 discharge rate which is competitive with experimental lithium/sulfur and lithium/oxygen cells.

20

Acknowledgements

- We would like to thank Mr. Max Tsao (UDRI) for his hard work on preparing and testing the cells.

21

401733 Foamed Mesoporous C/Si Composite Nanofibers Anode for Lithium Ion Batteries

Abstract:

A new porous composite nanofiber manufacturing route, combining electrospinning and foaming processes, was developed. Aluminum acetylacetonate was dissolved in a polyacrylonitrile (PAN)/ silicon nanoparticles suspension solution as the foaming agent. PAN composite nanofibers were first produced by an electrospinning process with mesopores generated via a sublimation-based foaming process during later thermal stabilization treatment. After carbonization, the obtained mesoporous carbon/silicon composite nanofiber mats were explored as the anode material of lithium ion batteries. Within this composite anode, mesopores provide needed buffering space to accommodate the large volume expansion and consequent stress of silicon during lithiation. This effectively mitigated silicon pulverization issue and helped achieve higher reversible capacity and better capacity retention in later battery tests when compared with anodes made of nonporous composites nanofibers and carbon nanofibers alone.

401598 Transport in Constrained Polymer-Electrolyte Membranes

Abstract:

Water transport in polymer-electrolyte membrane fuel cells is known to play a vital role in the efficient operation of these devices. The polymer-electrolyte membrane must be hydrated, to facilitate conduction of hydrogen ions; water formed as a by-product of the cell reaction at the cathode needs to be removed to avoid flooding. Moreover, PEMFCs operate in conditions where the membrane is not allowed to swell freely upon water uptake. This physical constraint can generate a stress differential in the membrane associated with water flux.

Water transport in the membrane is commonly acknowledged to occur by molecular diffusion, pressure diffusion, or electro-osmotic drag [1]; at higher water contents, external stress or pressure gradients across the membrane can also cause fluid flow directly by hydraulic permeation [2]. Most mathematical models, as well as experiments, treat the diffusive and hydraulic-permeation effects in isolation. There is no inherent contradiction in establishing a system where both these forces act in parallel, however. In addition to charge and material balances, we have added a local momentum balance that couples stress-induced compression and flow (permeation) into the transport system, as well as a thermodynamic equation of state that accounts for the local distribution of swelling incurred by variations in water content. Also, we argue that vapor-equilibrated and liquid-equilibrated cases can be handled by a single set of transport laws.

This talk will focus on reconciling all modes of transport in a single, thermodynamically rigorous macroscopic transport model accounting for simultaneous charge, water, and momentum transport in ionomer membranes. Composition and pressure dependences of properties such as activity coefficients, membrane compressibility and partial molar volumes will be explored from the perspective of thermodynamic constraints, and will be elucidated using experimental data from the literature.

The results help to rationalize phenomena associated with membrane swelling upon water sorption, as well as the consequences of constraining a membrane physically as water flux occurs. A comprehensive model will not only depict more accurate water profiles in the membrane, which can aid the design of effective water management schemes for fuel cells, but can also inform future modeling of swelling-induced membrane fatigue.

1. Weber A.Z., & Newman J. (2004). Transport in Polymer-Electrolyte Membranes. II. Mathematical Model. *Journal of The Electrochemical Society*. *151(2)*, A311-A325.

2. Duan Q., Wang H., & Benziger J. (2012). Transport of Liquid Water through Nafion Membranes. *Journal of Membrane Science*. *392-393*, 88-94.

ADVANCES IN THE LOCKHART-MARTINELLI APPROACH TO PREDICT PRESSURE DROP IN PEM FUEL CELLS

Paper number 400958

Dr. Ryan Anderson, Montana State University, Bozeman, MT, USA

Dr. Lifeng Zhang, University of Saskatchewan, Saskatoon, Canada

ABSTRACT

The water produced in proton exchange membrane fuel cells can lead to performance reductions if not properly managed. Complex two-phase flows can emerge throughout the cell, including in the gas delivery channels where the two-phase system causes an increase in the pressure drop. The Lockhart-Martinelli (LM) approach to predict two-phase pressure drop has been updated previously for the unique water introduction method in active fuel cells where the water emerges from a gas diffusion layer perpendicular to the direction of gas flow. Recent work has studied the Chisholm parameter C in terms of key fuel cell operating variables (e.g. relative humidity, temperature, materials, gas stoichiometry). However, the results of the analysis are unclear. A new method is proposed by which C is determined as a function of the operating current density and the flow regime. To complete this analysis, a new flow regime map is presented with an accumulating flow regime in addition to the traditional fuel cell flow regimes of single-phase, film/droplet, and slug. An additional force balance is presented to corroborate the superficial gas and liquid velocity bounds of the accumulating regime. These results allow a fuel cell designer to quickly judge the impact of design and operating conditions on two-phase flow pressure drop.

INTRODUCTION

Proton exchange membrane (PEM) fuel cells are under study as a source of clean energy for the future. Key benefits include high power density and no emissions at the point of use. The byproducts of the electrochemical reaction between oxygen and hydrogen are electricity, heat, and water. However, managing the water produced in the reaction remains a challenge due to competing effects: too little water dehydrates the membrane and too much water 'floods' the fuel cell's components (catalyst layer, gas diffusion layer (GDL), flow channel). Improper water management reduces cell performance [1,2], and there is a narrow range between dehydration and flooding where optimum performance is found. Usually, attempts are made to manage the water by modifying the operating conditions and/or redesigning components [3,4].

A specific area of research in PEM fuel cell water management is the multiphase (gas-liquid) pressure drop prediction [5], as the pressure drop represents a parasitic power loss. A simple, approximate method to determine the multiphase pressure drop is with the approach of Lockhart and Martinelli (LM) [6,7]:

$$\varphi_g^2 = 1 + C\chi + \chi^2 \tag{1}$$

Here, φ_g^2 is the two-phase flow multiplier, which is the ratio of the gas-liquid pressure drop to the single-phase gas pressure drop, and C is the Chisholm parameter. In the case of laminar-laminar flow, as typically found in fuel cells, χ^2 simplifies to a ratio of the single-phase pressure drops:

$$\chi^2 = \frac{\Delta P_l}{\Delta P_g} = \frac{u_L \mu_L}{u_g \mu_g} \tag{2}$$

However, water does not enter the flow channel in PEM fuel cells in the same manner in which the equations were originally developed. The traditional water injection, for which the LM approach was developed, assumes the water and air mix before entering the flow channel (Figure 1a Case a). However, in PEM fuel cells water enters the flow channel through a porous 'gas diffusion layer' perpendicular to the direction of gas flow. From this perspective, two scenarios could thus occur: uniform water injection where the same amount of water is injected at every channel location (Figure 1a Case b) or non-uniform water injection where more water enters at the channels' outlet (Figure 1a Case c). For parallel channels, a linear pressure drop is typical, and it can be inferred that more water will break through the porous media where the channel pressure is lower. This assumption leads to a linear water injection profile, which is a realistic injection scheme in PEM fuel cells. The non-uniform injection scheme therefore leads to a parabolic profile of the liquid water velocity down the length of the channel. Zhang et al. [8] utilized this relationship for liquid velocity as a function of channel length and integrated the original LM expression, leading to a two-phase flow multiplier prediction specific to PEM fuel cells:

$$\varphi_g^2 = 1 + \left(\frac{1}{2}\right) C\chi + \left(\frac{1}{3}\right)\chi^2 \tag{3}$$

Recent work has led to multiple research groups presenting two-phase flow multipliers for active PEM fuel cells. The literature provided the data sets for the analysis in this paper [9-11]. A collection of that available data is shown in Figure 1b.

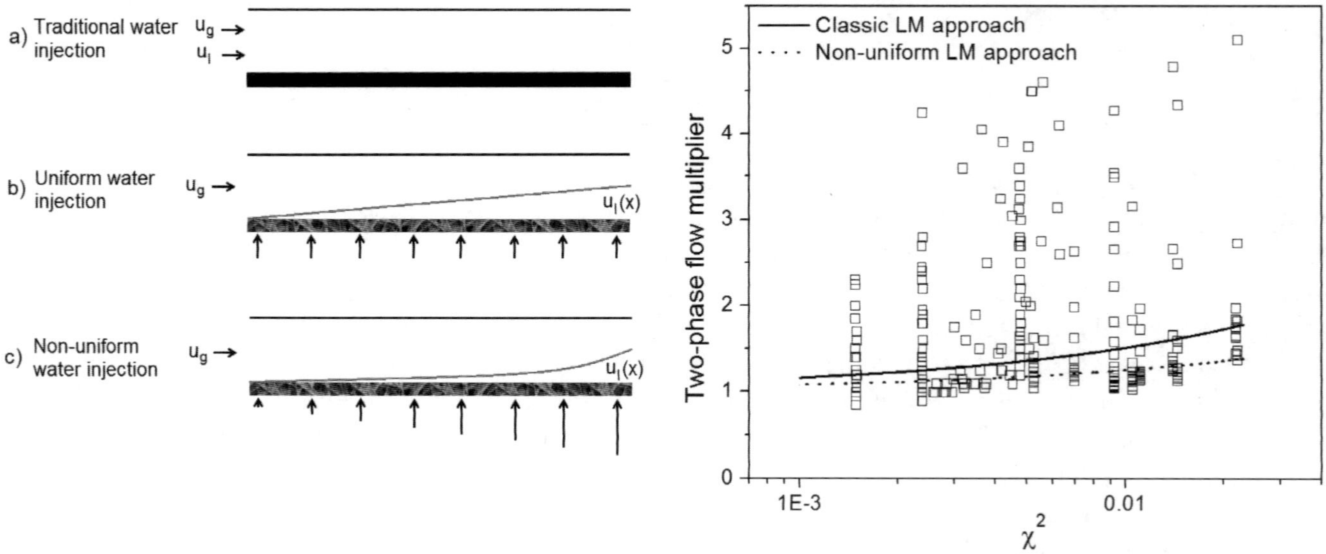

a) b)

Figure 1. a) Two-phase flow water injection schemes [with permission from Elsevier] b) Available literature data on two-phase flow multiplier from references [9-11]. The predictions for the Classic and Non-uniform LM approach come from Equations 1 and 3, respectively.

The datasets are described thoroughly in [12,13]. The data in [9-11] are all from electrochemically active PEM fuel cells and utilize the parallel channel flow field configuration. The velocities needed for χ^2 are determined from the current density, gas stoichiometry, channel design, temperature, and pressure, while the viscosities are calculated as a function of temperature and relative humidity. For typical laminar-laminar flow, a value of C = 5 is used to show the predicted results from Eqn. 1 (classic/traditional LM approach) and Eqn. 3 (non-uniform LM approach). It is clear from Figure 1b neither expression adequately predicts the experimentally measured two-phase flow multipliers in active PEM fuel cells.

To account for this, our work has focused on C as a function of 1) two-phase flow regime and 2) current density (air and water flow rate). This paper highlights recent contributions related to applying the Lockhart-Martinelli approach to low temperature PEM fuel cells. Firstly, an overview of the accumulating flow regime is provided with a particular emphasis on its relationship to pressure drop modeling. Then, results of a force balance study are explained to corroborate the 'location' of the accumulating regime on the PEM fuel cell flow regime map. Initial computational works are finally discussed to highlight the prediction of the accumulating flow regime and other key regimes in PEM fuel cells (slug, droplet/film, and mist/single-phase).

NOMENCLATURE

$2B$	channel height (m)	We	Weber Number
a_{Cho}	constant from Cho et al. [21]	φ_g^2	two-phase flow multiplier
b_{Cho}	constant from Cho et al. [21]	γ	surface tension (N m^{-1})
C	Chisholm parameter	χ^2	Martinelli parameter
ΔP	pressure drop (Pa)	μ	viscosity (Pa s)
R_0	droplet radius (m)	θ_s	static contact angle (^{o})
u_g	superficial gas velocity (m s^{-1})	θ_a	advancing contact angle (^{o})
u_L	superficial liquid velocity (m s^{-1})	θ_r	receding contact angle (^{o})
Re	Reynolds number		

RESULTS AND DISCUSSION

Our recent work has led to two key results: 1) the accumulating flow regime should be considered along with slug, film/droplet, and single-phase/mist flows in PEM fuel cells, and 2) the Chisholm parameter C is a function of both the flow regime (based on superficial velocities) and the current density. The results and discussion focus on the accumulating flow regime in PEM fuel cells, a force balance analysis to define when a droplet will remain adhered to the hydrophobic channel surface, and a computational fluid dynamics (CFD) approach to predict all relevant flow regimes in this multiphase system.

Accumulating flow regime in PEM fuel cells

By analyzing the published data in [9-11], the accumulating flow regime in PEM fuel cells was proposed for the first time [13], and these results are shown in Figure 2. Determining the bounds for the accumulating regime was an iterative procedure. Firstly, χ^2 was determined for all groups at the specific operating conditions. Then, C was solved for in Eqn. 3 knowing the experimentally reported two-phase flow multiplier and the calculated value of χ^2. Various values of C were then considered to help define a threshold where the two-phase flow multiplier would be considered 'too high' for the traditional approaches. Values included C = 5, 10, 15, 25, and 50. These data were then plotted in a flow regime map as a function of superficial gas and liquid velocities. The regimes for slug, film/droplet, and single-phase/mist were kept the same as those proposed by Hussaini and Wang [9]. Various boundaries were then proposed for the superficial gas and liquid velocities for an accumulating regime. With these four regimes defined/proposed, the data was separated, and each flow regime was plotted as a function of the current density (a key fuel cell variable). Based on these relationships, boundaries for the accumulating regimes and a threshold for *C* could be determined because certain *C* vs. current density relationships were expected. For instance, in single-phase flow, *C* should be low and not vary strongly with current density. Conversely, slug flow and accumulated water would cause a much higher pressure drop, so higher values of *C* were expected.

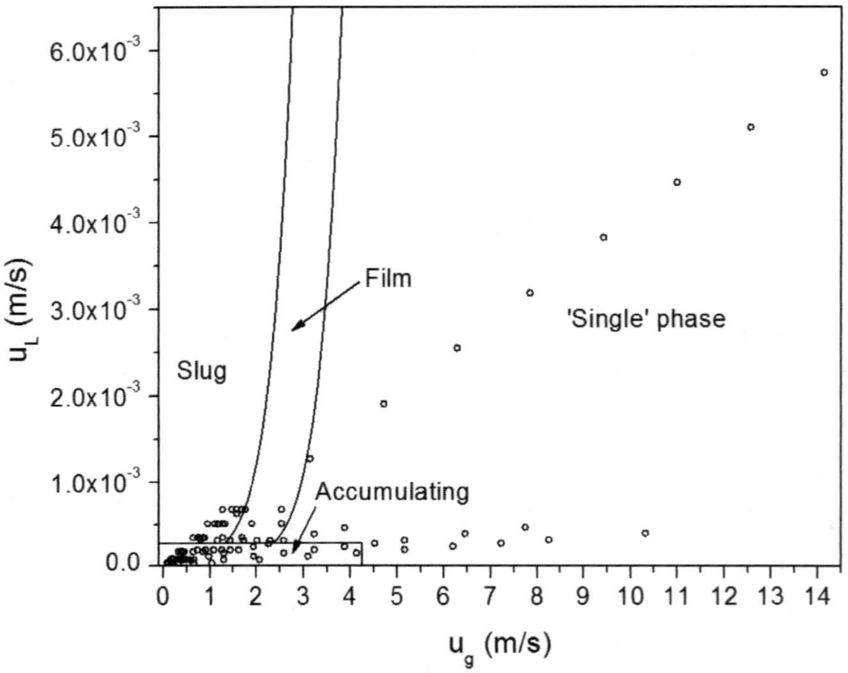

Figure 2. Flow regime map from [13] showing the approximate regions of the slug, film, single-phase, and accumulating flow regimes [with permission from Elsevier]

After this procedure was repeated several times, the accumulating region was proposed with $u_g <$ 4.25 m/s and $u_L < 2.75\text{x}10^{-4}$ m/s based on $C = 15$. This procedure was inherently qualitative but matches expected physics and visual observation of regimes. It is also worth noting that other methods to determine *C*, such as the approaches in minichannels [14,15], do not lead to good agreement. The need for the accumulating regime is supported by several groups that have noted long residence times of liquid water in the channels [16-19]. For example, Dunbar et al.'s [19] results with MRI show liquid water has residence times of ~175 seconds in some cases.

Corroboration with an analytical force balance approach

To further analyze the location of the accumulating flow regime in the flow field map in Figure 2, a force balance analysis was conducted to determine when a droplet was likely to detach from the gas diffusion layer surface. Adhered droplets support the accumulating flow regime. Much literature data is available on the water droplet dynamics in PEM fuel cells [20-26]. Our recent work [27] used the approach from Cho et al. [21] to relate detachment velocity and the droplet size for the operating conditions and designs in the data from [9-11]:

$$We_r = \left[\frac{4}{a_{Cho}} \frac{\pi \sin^2 \theta_s \sin(1/2)(\theta_a - \theta_r)}{(\theta_s - \sin \theta_s \cos \theta_s)} \right] \mathrm{Re}_g^{-b_{Cho}} \quad (4)$$

We_r is the Weber number, Re is the Reynolds number, and θ_s is the static contact angle. The advancing and receding contact angles are also considered. We and Re are defined as:

$$We = \frac{\rho u_g^2 R_0}{\gamma} \quad ; \quad \mathrm{Re}_g = \frac{\rho u_g (2B)}{\mu} \quad (5; 6)$$

Constants a_{Cho} and b_{Cho} are estimated based on the following equations [21]:

$$a_{Cho} = 46.27 \times \left(\frac{D}{(2B)} \right)^{0.1757} \quad ; \quad b_{Cho} = 0.2158 \times \left(\frac{D}{(2B)} \right) - 0.6384 \quad (7; 8)$$

All relevant conditions are in Table 1 of our recent work [13].

Figure 3 shows these results for the critical droplet diameters based on the conditions and geometries in [9-11]. Droplet sizes smaller than the critical values are considered stable, and will thus remain accumulated on the gas diffusion layer surface. Droplet sizes above the critical values are considered unstable, leading to droplet detachment from the surface.

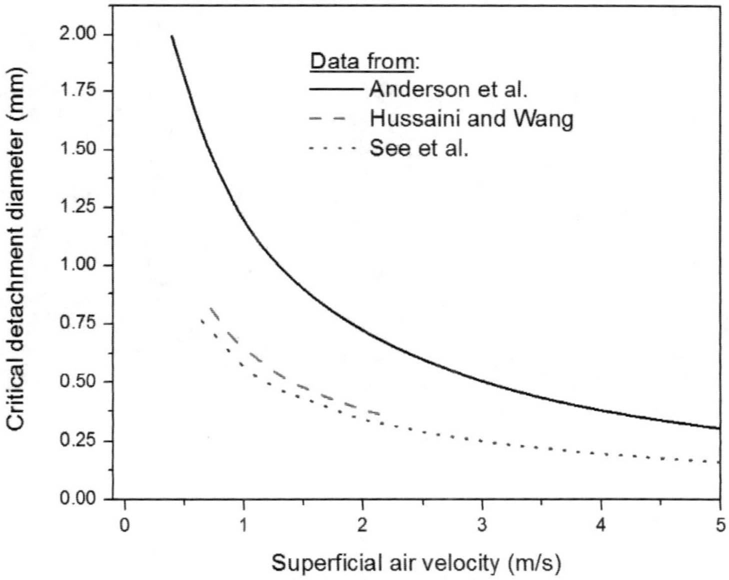

Figure 3. Predicted critical droplet sizes for Hussaini and Wang's [9], See's [11], and Anderson et al.'s [10] fuel cell. Specific conditions are in [13,27].

Specifically, this analysis only considers the droplet growing on the GDL surface with no influence from the wall. Depending on the specific flow channel geometry in the fuel cell, the droplet's critical diameter may be larger than the flow channel itself. In that case, a droplet would wet the hydrophilic channel wall and detach from the GDL.

These results can be qualitatively compared to the proposed accumulating flow regime in Figure 2 based on droplet stability. At low liquid velocities, smaller drops will form. Small droplets are in the Stable region, which requires a high superficial air velocity for detachment. This is consistent with the accumulating regime, where small droplets grow at low liquid velocities and remain pinned on the GDL surface. At low gas velocities, the liquid velocity can eventually lead to larger droplets, which can fall into the unstable region and become slug flow or film flow. These results show good qualitative agreement with those in Figure 2.

CFD VOF approach: Current results and future work

The force balance based analysis here only considers a single droplet centred in a flow channel. However, several other possibilities exist for how the droplet can move in the flow channel. For instance, a detached droplet can combine with other downstream droplets. In reality there is a distribution of pores from which droplets emerge. Also, the wetting properties of the channel sidewalls are hydrophilic. To account for all of these variables, a CFD approach will be taken to understand the predicted flow patterns in the fuel cell. Work will focus on the four main regimes reported in Figure 2. In-house code with the volume of fluid (VOF) method is currently being implemented to complement the results presented here[1]. An example of initial work with a single-droplet growing from one pore is shown in Figure 4.

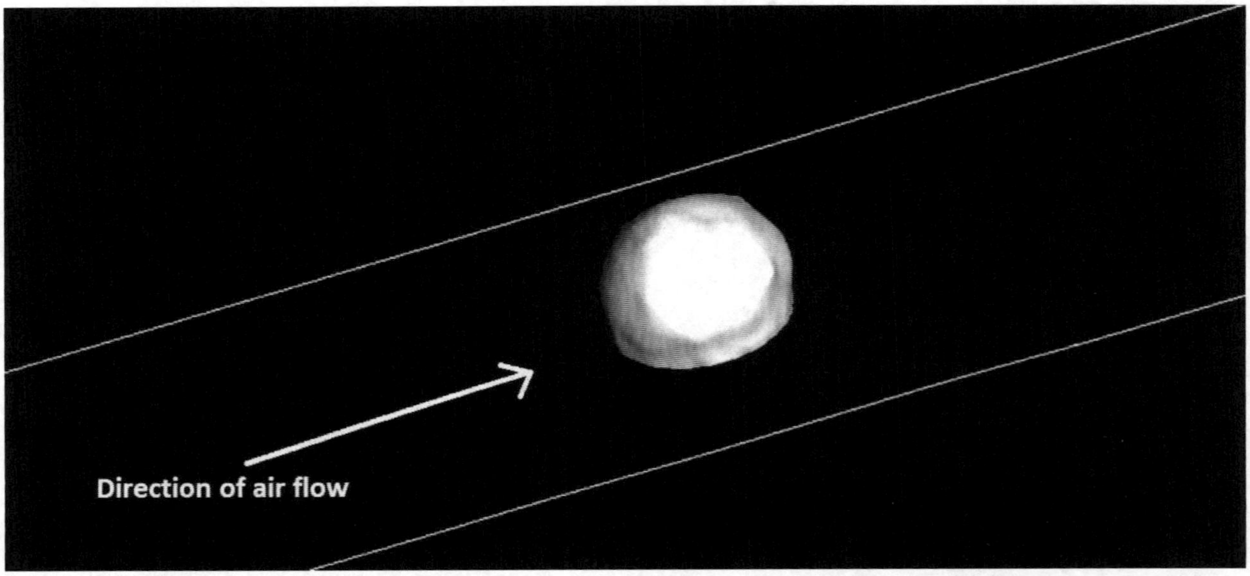

Figure 4. Three-dimensional computational domain of an air and water multiphase system, with a single droplet emerging from a pore

CONCLUSIONS

This paper summarizes recent contributions to predicting the two-phase pressure drop in PEM fuel cells via an updated Lockhart-Martinelli approach. To do so, an accumulating flow regime is firstly presented in addition to the previously accepted fuel cell flow regimes of slug, droplet/film, and single-phase/mist. A force balance analysis highlights how the location of the accumulating flow regime is qualitatively reasonable. Future work will consider a force balance analysis to determine slug flow and film flow. A CFD VOF approach will complement that work.

[1] In collaboration with Dr. Mark Owkes and Eric Cauble at Montana State University

ACKNOWLEDGMENTS

We would like to thank Erica Eggleton for analyzing the data in the accumulating flow regime. Thanks are also given to the University of Saskatchewan and Montana State University.

REFERENCES

[1] Li, H., Tang, Y., Wang, Z., Shi, Z., Wu, S., Song, D., ... & Mazza, A. (2008). A review of water flooding issues in the proton exchange membrane fuel cell. *Journal of Power Sources*, 178(1), 103-117.

[2] Dai, W., Wang, H., Yuan, X. Z., Martin, J. J., Yang, D., Qiao, J., & Ma, J. (2009). A review on water balance in the membrane electrode assembly of proton exchange membrane fuel cells. *International Journal of Hydrogen Energy*, 34(23), 9461-9478.

[3] Blanco, M. (2011). Study of selected water management strategies for proton exchange membrane fuel cells. *Ph.D. thesis*, University of British Columbia, Vancouver, BC.

[4] Anderson, R., Zhang, L., Ding, Y., Blanco, M., Bi, X., & Wilkinson, D. P. (2010). A critical review of two-phase flow in gas flow channels of proton exchange membrane fuel cells. *Journal of Power Sources*, 195(15), 4531-4553.

[5] Liu, X., Guo, H., Ye, F., & Ma, C. F. (2007). Water flooding and pressure drop characteristics in flow channels of proton exchange membrane fuel cells. *Electrochimica Acta*, 52(11), 3607-3614.

[6] Lockhart, R. W., & Martinelli, R. C. (1949). Proposed correlation of data for isothermal two-phase, two-component flow in pipes. *Chem. Eng. Prog*, 45(1), 39-48.

[7] Chisholm, D. (1967). A theoretical basis for the Lockhart-Martinelli correlation for two-phase flow. *International Journal of Heat and Mass Transfer*, 10(12), 1767-1778.

[8] Zhang, L., Bi, X. T., Wilkinson, D. P., Anderson, R., Stumper, J., & Wang, H. (2011). Gas–liquid two-phase flow behavior in minichannels bounded with a permeable wall. *Chemical Engineering Science*, 66(14), 3377-3385.

[9] Hussaini, I. S., & Wang, C. Y. (2009). Visualization and quantification of cathode channel flooding in PEM fuel cells. *Journal of Power Sources*, 187(2), 444-451.

[10] Anderson, R., Wilkinson, D. P., Bi, X., & Zhang, L. (2011). Two-phase flow pressure drop hysteresis in an operating proton exchange membrane fuel cell. *Journal of Power Sources*, 196(19), 8031-8040.

[11] See, E. (2013). In situ measurement, characterization, and modeling of two-phase pressure drop incorporating local water saturation in PEMFC gas channels. *M.S. thesis*, Rochester Institute of Technology, Troy, NY.

[12] Anderson, R., Zhang, L., Wilkinson, D.P. (2014). An analysis of two-phase flow pressure drop in operating proton exchange membrane fuel cell channels with the Lockhart-Martinelli approach. *12th International Conference on Nanochannels, Microchannels, and Minichannels*, FEDSM2014-21288.

[13] Anderson, R., Eggleton, E., & Zhang, L. (2015). Development of two-phase flow regime specific pressure drop models for proton exchange membrane fuel cells. *International Journal of Hydrogen Energy*, 40(2), 1173-1185.

[14] English, N. J., & Kandlikar, S. G. (2006). An experimental investigation into the effect of surfactants on air-water two-phase flow in minichannels. *Heat Transfer Engineering*, *27*(4), 99-109.

[15] Lee, C. Y., & Lee, S. Y. (2008). Pressure drop of two-phase plug flow in round mini-channels: influence of surface wettability. *Experimental Thermal and Fluid Science*, 32(8), 1716-1722.

[16] Lu, Z., Kandlikar, S. G., Rath, C., Grimm, M., Domigan, W., White, A. D. & Trabold, T. A. (2009). Water management studies in PEM fuel cells, Part II: Ex situ investigation of flow maldistribution, pressure drop and two-phase flow pattern in gas channels. *International Journal of Hydrogen Energy*, 34(8), 3445-3456.

[17] Carton, J. G., Lawlor, V., Olabi, A. G., Hochenauer, C., & Zauner, G. (2012). Water droplet accumulation and motion in PEM (Proton Exchange Membrane) fuel cell mini-channels. *Energy*, 39(1), 63-73.

[18] Hellstern, T., Gauthier, E., Cheah, M. J., & Benziger, J. B. (2013). The role of the gas diffusion layer on slug formation in gas flow channels of fuel cells. *International Journal of Hydrogen Energy*, 38(35), 15414-15427.

[19] Dunbar, Z. W., & Masel, R. I. (2008). Magnetic resonance imaging investigation of water accumulation and transport in graphite flow fields in a polymer electrolyte membrane fuel cell: Do defects control transport?. Journal of Power Sources, 182(1), 76-82.

[20] Cho, S. C., Wang, Y., & Chen, K. S. (2012). Droplet dynamics in a polymer electrolyte fuel cell gas flow channel: Forces, deformation, and detachment. I: Theoretical and numerical analyses. *Journal of Power Sources*, 206, 119-128.

[21] Cho, S. C., Wang, Y., & Chen, K. S. (2012). Droplet dynamics in a polymer electrolyte fuel cell gas flow channel: Forces, Deformation and detachment. II: Comparisons of analytical solution with numerical and experimental results. *Journal of Power Sources*, 210, 191-197.

[22] Kimball, E., Whitaker, T., Kevrekidis, Y. G., & Benziger, J. B. (2008). Drops, slugs, and flooding in polymer electrolyte membrane fuel cells. *AIChE Journal*, 54(5), 1313-1332.

[23] Kumbur, E. C., Sharp, K. V., & Mench, M. M. (2006). Liquid droplet behavior and instability in a polymer electrolyte fuel cell flow channel. *Journal of Power Sources*, 161(1), 333-345.

[24] Lee, S. K., & Ito, K. (2014). Cross-sectional visualization and analysis of droplet behavior in gas flow channel in PEFC. *Journal of The Electrochemical Society*, 161(1), F58-F66.

[25] Schillberg, C., Kandlikar, S., A review of modes for water droplet detachment from the gas diffusion later-gas flow interface in PEMFCs, *Proceedings of the Fifth International Conference on Nanochannels, Microchannels and Minichannels*, Puebla, Mexico, June, 2007, pp. 299-310.

[26] Fang, C., Hidrovo, C., Wang, F. M., Eaton, J., & Goodson, K. (2008). 3-D numerical simulation of contact angle hysteresis for microscale two phase flow. *International Journal of Multiphase Flow*, 34(7), 690-705.

[27] Anderson, R., Zhang, L., An updated two-phase flow regime map in active PEM fuel cells based on a force balance approach, *Proceedings of the ASME 2015 13th International Conference on Nanochannels, Microchannels, and Minichannels*, July 6-9, 2015, San Francisco, California, USA, InterPACKICNMM2015-48527.

399349 Intermediate Temperature Solid Oxide Fuel Cell Based on Proton-Conducting Ceramic Electrolyte

Abstract:

Fuel cells cleanly and efficiently convert chemical energy directly to electrical energy. With highly efficient energy conversion and very low greenhouse gas emissions, fuel cells can play a pivotal role in reducing our nation's dependence on foreign oil while lessening the emission of greenhouse gases. To now, however, material issues have impeded the commercialization of fuel cells. An oxygen-ion conducting electrolyte, such as yttria-stabilized zirconia, has been traditionally used for the fabrication of solid oxide fuel cells (SOFCs). Due to the high activation energy for conduction, SOFCs based on stabilized zirconia are normally operated at temperatures approaching 1000°C. This high operating temperature leads to the corrosion and incompatibility of materials, severely limits the selection of materials that might be suitable as the electrical interconnect, and increases the difficulty of forming seals and manifolding. To overcome these problems, efforts are underway to develop electrolytes that can operate at intermediate temperatures. Proton-conducting oxides have attracted considerable attention as an electrolyte for SOFCs that operate at intermediate-temperature (≈600°C). Use of a proton-conducting electrolyte in SOFCs eliminates the dilution of fuel by water, which occurs at the anode in conventional SOFCs based on an oxygen-ion-conducting electrolyte. We have fabricated a SOFC using a $BaCe_{0.8}Y_{0.2}O_x$ (BCY) proton conductor as the electrolyte. A dense BCY film (≈10-μm thick) was deposited on a porous Ni/BCY cermet (i.e., ceramic/metal composite) substrate by a dip-coating process. The gas-permeable Ni/BCY cermet substrate backed with nickel mesh was used as the anode, and platinum paste backed with platinum mesh served as the cathode. The current-voltage characteristics of the BCY-based SOFC were measured in the temperature range 450-800°C using wet air on the cathode side and hydrogen on the anode side. The open-circuit voltage was close to the theoretical value at lower temperatures (<600°C) and was about 85% of the theoretical value at 800°C. The lower measured voltages at high temperature may be due to the development of electronic conductivity in the BCY electrolyte. Gas leakage due to an incomplete seal may also contribute to the deviation from the theoretical voltage at high temperature. The peak power density of the fuel cell was ≈90 and ≈1500 mW/cm^2 at 450 and 800°C, respectively. The power density increased as temperature increased because the cell resistance decreased. Our efforts to develop SOFCs based on proton conductors will be summarized in this talk.

Work supported in part by the U.S. Department of Energy, Advanced Research Projects Agency – Energy (ARPA-E) under Contract DE-AC02-06CH11357.

PROCESS DEVELOPMENT OF MANGANESE-BASED OXYGEN CARRIERS FOR OXIDATIVE COUPLING OF METHANE IN A PRESSURIZED CHEMICAL LOOPING SYSTEM

Elena Y. Chung, The Ohio State University, Columbus, OH
William K. Wang, The Ohio State University, Columbus, OH
Hussein Alkhatib, The Ohio State University, Columbus, OH
Sourabh Nadgouda, The Ohio State University, Columbus, OH
Michael A. Jindra, The Ohio State University, Columbus, OH
John A. Sofranko, EcoCatalytic Technologies, Monmouth Junction, NJ
Liang-Shih Fan, The Ohio State University, Columbus, OH

Keywords: chemical looping, oxidative coupling of methane, methane utilization

Abstract

With the recent growth in global shale gas production, interest has increased in upgrading shale gas and its primary component, methane, to high value-added products. However, a direct method of upgrading methane has been limited by methane's high molecular stability and process economics. Since the 1980s, oxidative coupling of methane (OCM) has demonstrated promise in producing higher hydrocarbons, specifically ethylene and ethane (C_2 products). Typical OCM schemes catalytically convert methane to C_2 products using gaseous oxidants, which can lower process efficiencies by requiring energy-intensive air separation operations. In order to overcome these process limitations, the chemical looping technology platform presents an attractive opportunity to convert hydrocarbons to flexible products utilizing an intermediate oxygen carrier rather than gaseous co-fed oxidants.

This study investigates the application of the chemical looping approach for OCM to selectively control the redox reactions. Two reducible manganese-based oxidative catalysts or oxygen carriers were synthesized and compared. Specifically, parametric co-feed or redox experiments were conducted in a high-pressure, fixed bed reactor with pressures between atmospheric and 507 kPa, at 850°C and a 2.7:1 steam-to-methane ratio. The effects on the oxidation and reduction reactions are also investigated in a thermogravimetric analyzer (TGA). The effects of high pressure on the oxygen carriers were characterized by x-ray diffraction (XRD). The experimental results are coupled with process simulations of a theoretical commercial OCM co-feed or redox chemical looping system for either ethylene or liquid fuel production. Experimental results demonstrate the effects of pressure and steam on methane conversion and C_2+ selectivity with the OCM chemical looping process. With proper heat management, the OCM chemical looping process simulations present promising reactor and process designs for an alternative direct scheme for methane utilization.

Introduction

Methane and other carbon-based feedstocks can be partially oxidized to chemicals indirectly (via syngas) or directly. A direct approach to monetizing methane would potentially simplify and reduce chemical processing costs.[1] One promising direct method of converting methane to ethylene is oxidative coupling of methane (OCM), which was first proposed by Keller and Bhasin in 1982.[2] However, this reaction system has a challenging trade-off relationship between achieving high methane conversion and high ethane and ethylene (C_2) selectivity. During high methane conversion, partial oxidation of methane thermodynamically favors undesirable and competing combustion products such as carbon dioxide and carbon monoxide, which reduces the C_2 selectivity. Though proposed in the 1980s, the low yields of OCM have prevented the process scale-up to commercial use. It is suggested that a total C_2 yield of over 25%[2,3] or 35-40%[11] would allow the OCM process to be economically feasible.

Based on the reactant gas supply, OCM has been explored by two approaches: (1) simultaneous[2] or co-feeding of methane and oxygen (or air), which can also be classified as the pure catalytic or co-feed OCM; and (2) sequential[2], cyclic or periodic (or alternating) feeding of methane individually from air, which can also be classified as reduction-oxidation (redox) OCM. Figure 1 and Figure 2 show the two reaction operating approaches for the OCM reaction.

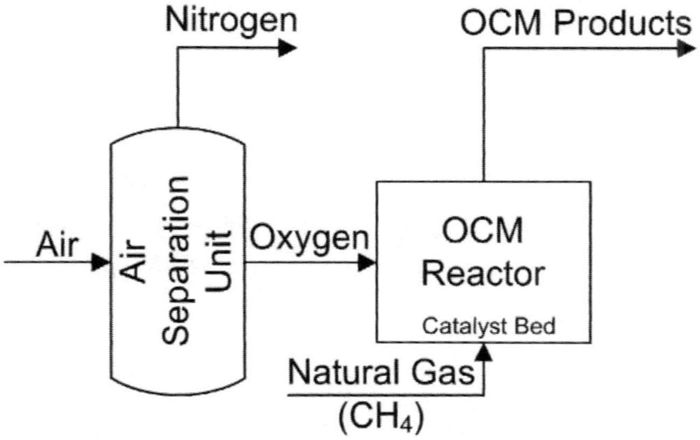

Figure 1: Co-feed mode using molecular oxygen and catalyst.

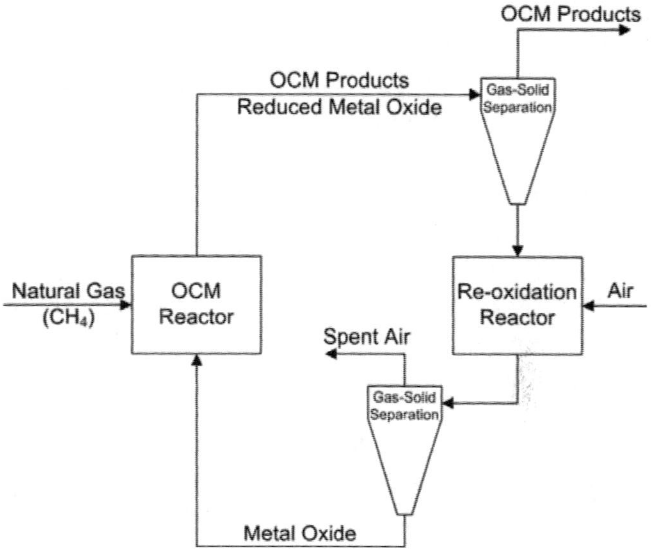

Figure 2: Redox mode using metal oxide with catalytic properties.

Since the pioneering research into OCM[2,8], hundreds of catalytic materials have been studied for OCM. Promising catalysts include alkali and alkaline metal-supported basic oxides and transition metal oxides.[5,9] Manganese oxides have demonstrated its promise as catalysts for OCM.[4,9] Since the 1990s, a sodium-promoted manganese oxide ($Mn/Na_2WO_4/SiO_2$) from Fang et al.[9] has been a popular co-feed OCM catalyst for its high stability and single-pass C_2 product yield of 25-30%. Reducible manganese oxides have been used for redox OCM with C_2 yields of 20%.[2,4,5]

The OCM redox operation mode and the reducible metal oxide catalysts are conceptually analogous to the redox cycles and reducible oxygen carriers of the oxy-fuel combustion process of chemical looping. Recently studied for CO_2 capture of solid or gaseous fuel combustion, chemical looping uses the lattice oxygen from an oxygen carrier that is circulated between a reducer (or fuel reactor) and a combustor (or air reactor) to transfer oxygen.[6,7] Chemical looping avoids direct contact between gaseous oxygen and the fuel such as methane, thereby reducing the need for air separation units (ASUs) and allowing for inherent separation of CO_2. Thus, the chemical looping process provides another approach to OCM. This paper compares the potential of co-feed OCM and redox OCM experimentally with pressurized fixed bed studies and conceptually with process modeling for the production of ethylene or liquid fuels.

OCM Experimentation

Oxygen Carrier Preparations

Two reducible manganese-based oxidative catalysts or oxygen carriers were synthesized and compared. The $Mn/Na_2WO_4/SiO_2$ catalyst was prepared following the mixture slurry method, which has demonstrated high activity and uniform surface and bulk distribution.[10-13] Silica gel (Davisil Grade 633, 200-425 mesh) was mixed with distilled water at 105ºC. An aqueous solution of sodium tungstate dihydrate (Sigma-Aldrich, ACS reagent, 99% pure) and manganese (II) nitrate hydrate (Sigma-Aldrich, 98% pure) was added dropwise with a molar ratio of 2.14:1:91 $Mn(NO_3)_2$:Na_2WO_4:SiO_2 in order to obtained the optimized content of 2 wt% Mn with 5 wt% Na_2WO_4 on SiO_2. The aqueous solution was dried at 105ºC in air for 24 h. The final solids were calcined at 800ºC in air for 5 h.

The $Mn/Na_2WO_4/SiO_2$ catalyst is compared to a proprietary manganese-based oxidative catalyst supplied from EcoCatalytic Technologies. The manganese-based oxygen carriers were sieved to a particle diameter of 53 to 74 μm.

Oxygen Carrier Characterizations

Prior to fixed bed experiments, the catalyst synthesis method was verified for reproducibility with results from literature. X-ray diffraction (XRD) patterns of the catalysts were analyzed by a Rigaku SmartLab High Resolution X-ray diffractometer equipped with Cu Kα radiation and a diffraction beam monochromator and a scintillation detector. The XRD analysis confirmed phases of cristobalite, quartz, Na_2WO_4 and Mn_2O_3 as expected from literature.[10-12]

The oxidative catalysts were tested for its oxygen carrying capacity using a Setaram Setsys Evolution thermogravimetric analyzer (TGA) at a heating ramp rate of 50 ºC/min under air (21% O_2 balanced by N_2). At 850ºC, approximately 20 mg of prepared OCM catalysts underwent three steps: (1) oxidization for four hours under air; (2) inert purge for four hours under nitrogen; and (3) reduction for four hours under 37.5% ultra-high purity H_2 balanced by N_2. Hydrogen gas was used to avoid any potential carbon deposition. All experiments were conducted with a gas flow of 300 ml_n/min or more (normalized to 1 bar and 0ºC) to minimize mass transfer diffusion effects. The oxygen capacity was calculated for the $Mn/Na_2WO_4/SiO_2$ of 1.5% and for the proprietary catalyst of 6.0%.

Pressurized Fixed Bed Reactor Experiments with Steam

Depending on co-feed or redox mode, reactant feed gases of air, nitrogen and methane were mixed in a custom reactant gas-mixing manifold with DAQ Factory-controlled solenoid two-way and three-way valves. Desired reactant gas flow rates were set by FlowVision-controlled Alicat Scientific mass flow controllers.

The mixed reactant gases are fed to the fixed bed reactor system holding the oxygen carriers. A CoorsTek Ad-998 alumina fixed bed reactor (ID =12.7 mm) is used to test between 2-10 grams of oxygen carriers depending on oxygen capacity. The oxygen carriers were held in position with quartz wool. The reactor dead volume was filled with inert quartz particles of diameter 3 mm to minimize any secondary gas-phase reactions that can occur in the reactor void space. A back-pressure regulator was used to set the pressure of reactor.

The effluent from the reactor was sent through a gas conditioning system to remove moisture in order to protect the gas analysis instruments. Reactant gas and effluent analysis was conducted by online gas analysis by a Varian Micro-GC CP-4900 equipped with four columns to separate and quantify hydrogen, oxygen nitrogen, methane, carbon monoxide, butane, hexane, heptane, carbon dioxide, ethylene, ethane, acetylene and propane. Steam was injected during methane addition at a steam:CH_4 ratio of 2.67:1 in order to reduce carbon deposition.

Pressurized Fixed Bed Reactor in Co-Feed Mode

For co-feed experiments, the fixed bed reactor was heated to 850°C under air and nitrogen at 10°C/min. Prior to reaction temperature, the reactor was flushed for 15 minutes with pure nitrogen. At reaction temperature, a CH_4/O_2 ratio of 2.2 was co-fed into the reactor. These conditions were selected in order to verify the catalyst preparation method produced conversions and C_2 selectivity that are comparable to literature. Based on Dedov et al., it was expected to obtain C_2+ yields of 18% under the following conditions without steam injection.[10]

Pressurized Fixed Bed Reactor in Chemical Looping Redox Mode

For chemical looping redox experiments, the fixed bed reactor was heated to 850°C under air and nitrogen at 10°C/min. Prior to reaction temperature, the reactor was flushed for 15 minutes with pure nitrogen. At reaction temperature, a cyclic redox sequence was initiated by reduction with CH_4 at WHSV = 7.04 h^{-1} for 10 min and oxidized by air for 20 minutes where the reduction and oxidation steps were flushed for 10 minutes by pure nitrogen.

Results of Pressurized Fixed Bed Reactor Experiments with Steam

Table 1 and 2 are the compiled results from fixed bed reactor experiments at varying pressures for the two different oxygen carriers. Co-feed data are the averaged values. Redox data are collected from the maximum C_2 yield during reduction.

Table 1. Methane Conversion and C_2 Selectivity for $Mn/Na_2WO_4/SiO_2$

Reaction	**Co-feed Mode**			**Redox Mode**		
	CH_4 Conversion	*C_2 Selectivity*	*C_2 Yield*	*CH_4 Conversion*	*C_2 Selectivity*	*C_2 Yield*
Pressure = 101.3 kPa	52.2%	51.3%	26.8%	8.7%	82.2%	7.2%
304 kPa	52.6%	39.2%	20.6%	7.3%	75.9%	5.5%
506.6 kPa	60.1%	23.6%	14.2%	6.2%	69.4%	4.3%

Table 2. Methane Conversion and C_2 Selectivity for Proprietary Catalyst

Reaction	**Co-feed Mode**			**Redox Mode**		
	CH_4 Conversion	*C_2 Selectivity*	*C_2 Yield*	*CH_4 Conversion*	*C_2 Selectivity*	*C_2 Yield*
Pressure = 101.3 kPa	43.8%	37.8%	16.4%	15.7%	87.5%	13.8%
304 kPa	48.2%	26.3%	12.6%	15.2%	79.5%	12.1%
506.6 kPa	55.1%	15.5%	8.3%	30.1%	31.0%	9.3%

In both Table 1 and 2, the results demonstrate that co-feed experiments with the $Mn/Na_2WO_4/SiO_2$ catalysts produce higher yields that are consistent with literature values. Under redox conditions, the proprietary catalysts demonstrate better C_2 yields with selectivity as high as 87.5%. The general trend with increasing pressure is that methane conversion increased and C_2 selectivity decreased. This is consistent with literature investigations that found that with higher pressure, C_2 products decrease due to the thermodynamic preference for CO to form CO_2 and the dominance of gas-phase reactions occurring pre- and post-oxygen carriers.[11] Thus, the $Mn/Na_2WO_4/SiO_2$ and the proprietary catalysts favor different OCM approaches.

OCM Process Development and Analysis

In order to further explore the potential of co-feed OCM and chemical looping OCM, process modeling is employed. The OCM process can be divided into two independent sections: reaction and processing. Within the reaction section, the OCM reactor can be operated in either a co-feed mode using a catalyst and molecular oxygen or a redox mode using a metal oxide with catalytic properties as illustrated in Figure 1 and 2.[14] The co-feed mode requires an ASU to generate the oxygen while the redox mode requires cycling between natural gas and air.[15]

While the desired primary product from the OCM reactor is ethylene, the ethylene concentration is low and requires separation. The OCM co-feed or redox process for ethylene production is currently not economically competitive with the conventional process of ethylene production via steam cracking; this has led to investigating coupling OCM with electricity generation, Fischer-Tropsch reaction, methane reforming, oxygenates production, and liquid fuels production.[16-23]

Process Considerations

For the reaction section, the OCM reactor always operates at a high temperature, and potentially at an elevated pressure. When operating in the co-feed mode, the oxygen is supplied from a cryogenic ASU. While the ASU is commercially available and fairly commonplace in industrial processes, there are two major drawbacks when integrated into the OCM process. First, the ASU is energetically intensive, consuming between 912 and 1270 J/g oxygen at 95% (volume) purity. The lower energy consumption is associated with lower outlet pressures. Second, the oxygen temperature exiting the ASU is between 13°C and 30°C, while the OCM reactor operates around 800°C.

Under co-feed mode, heating both the natural gas and oxygen to OCM reactor temperature is necessary. In the redox mode, the oxygen is supplied from an oxygen carrier that possesses catalytic activity towards ethylene production. This removes the need for an ASU and pre-heating of the oxygen as the oxygen carrier can be regenerated at a temperature equal to or greater than the OCM reactor. Further, the redox OCM reactor and re-oxidation reactor can be individually operated at different temperatures and pressures. However, cyclic reactivity must be maintained and particle attrition minimized. From a heat and process integration standpoint, the advantage of operating the OCM reactor in redox mode coupled with the rapid advancement in chemical looping technologies in recent years yields the potential of significant energy and cost reductions for the OCM process.

For the processing section, the standard separation process employed in steam cracking plants requires significant energy requirements. After the OCM reactor, the hydrocarbon products are cooled, compressed, purified, and cryogenically distilled in multiple trains.

The OCM process can be used as an intermediate to produce alternative products. Since the products of the OCM reactor are a mixture of hydrocarbons, the potential exists to convert the hydrocarbon rich mixture into high-value products. Two process models have been simulated to highlight the difference between co-feed mode for ethylene production and redox mode for liquid fuels production. These process model configurations were chosen based on the reactor outlet product distribution where co-feed produces mostly ethylene and redox produces a wide mixture of olefins.

Process Simulation Set-up

Aspen Plus® was used to model the OCM process. Only the OCM reactor products were modeled using expected conversions and product yields based on input parameters. The catalyst and reaction kinetics were omitted; for the redox mode, the redox reaction was limited to the oxygen donation reactions. The processing for the co-feed mode and the redox mode were ethylene production and liquid fuels, respectively. Table 3 shows the process model assumptions used to model the OCM process.

The co-feed mode for ethylene production operates at 405.3 kPa, which has been estimated to be the most economically optimal pressure to operate the OCM reactor.[13, 15, 16] The OCM reactor products are cooled in two stages from 840°C at the reactor outlet to 90°C and 40°C. The first stage of cooling is used to heat the natural gas and oxygen to 820°C prior to entering the OCM reactor. The OCM reaction products are then compressed with interstage cooling such that the compressor outlet temperature remains below 100°C to reduce polymerization. Since the OCM reactor operates at 400 kPa, the first two stages of a typical five-stage compression is not necessary. A caustic wash then removes the carbon dioxide between the first and second compression stage, and water is completely removed after the last stage of compression. The purified gas is then transported to the cold box section for cryogenic distillation to recover the ethylene.

Table 3. Modeling Parameters for OCM Process

	Reaction	
	Co-feed Mode	**Redox Mode**
Methane Flow (kmol/h)	1000	1000
Methane T, P (°C, kPa)	25, 3040	25, 3040
CH_4:O_2 mole ratio	4:1	1.93:1 (16% excess oxygen)
Steam: CH_4 mole ratio	--	1.2:1
O_2 Purity	95% (volume)	--
O_2 T, P (°C, kPa)	20, 405.3	--
OCM Reactor T, P (°C, kPa)	840, 405.3	850, 304
Re-oxidation Reactor	--	900, 101.3
OCM Reactor Model	RSTOIC	RSTOIC
CH_4 Conversion	20%	24.55%
Product Selectivity	Reference [25] average	Reference [26]
Ethylene	46%	43.75%
Ethane	14%	3.33%
CO_x	40%	24.55%
	Processing	
Compression Stages[27] (kPa)	911.9, 1824, 3546	557, 1064, 2027
Final Product	Ethylene	Liquid Fuels

The redox mode for ethylene production operates at 304 kPa. A 1.2:1 steam:CH_4 ratio is used based on available heat to generate the steam. The OCM reactor is autothermal, while the re-oxidation reactor is slightly exothermic at 900°C. The OCM reactions are exothermic, but the excess heat is used to heat the inlet steam and natural gas to the reactor temperature. The exothermic heat of reaction from the re-oxidation reactor can be used to operate a steam turbine to generate electricity. The gas-solid separation efficiency for each one is 98%. This allows for accounting of the attrited fines exiting the process while also allowing for additional fresh solids to enter the system. The hydrocarbon products exiting the OCM reactor are cooled in three stages to 119°C, 118°C, and 40°C to allow for pre-heating of the steam and air. The hydrocarbon products are then compressed to 2027 kPa in three stages with interstage cooling. The compressed hydrocarbon products are then delivered to an oligomerization reactor where they are converted into liquid fuels. The oligomerization reactor is exothermic, so the hydrocarbon products are not pre-heated prior to entering the oligomerization reactor.

Process Analysis

The OCM reactions for both cases are exothermic. In the co-feed mode, 16 kW_{th} is produced per mole of methane in the OCM reactor. While for the redox mode, the OCM reactor net heat duty is zero due to steam energy requirements. However, 6.5 kW_{th} is produced per mole of methane in the re-oxidation reactor for the redox mode, while 2.2 kW_e per mole of methane electricity input is necessary to operate the ASU.

The largest advantage of the redox mode over the co-feed mode is obtained in the processing section. For both ethylene production and liquid fuels production, the compressors were operated identically with a polytropic efficiency of 86% and mechanical efficiency of 98%, both of which are on the high end for efficiencies in an ethylene plant. The compressor electricity requirements for ethylene production from the co-feed mode is 2 kW_e per mole of methane as compared to 1.6 kW_e per mole of methane for liquid fuels production from the redox mode. Thus, producing liquid fuels from a redox mode requires 40% less electricity for compression and does not require electricity to operate an ASU when producing ethylene from a co-feed mode. Finally, the production of high purity ethylene using cryogenic distillation is highly energy intensive while liquid fuels production via oligomerization is exothermic at intermediate temperatures sufficient to produce heat and/or electricity.

Overall, the co-feed mode for ethylene production requires 4.2 kW_e per mole of methane while producing 16 kW_{th} per mole of methane. The redox mode for liquid fuels production only requires 1.6 kW_e per mole of methane while producing 6.5 kW_{th} per mole of methane from the re-oxidation reactor and additional thermal energy from the exothermic heat of reaction from the oligomerization reactor. The heat of reaction from the oligomerization reactor has not been quantified, as it will vary based on liquid fuels product distribution. Given the energy requirements for ethylene separation and the energy production from olefin oligomerization, it can be concluded that liquid fuels production is more favorable than ethylene production from an energy standpoint. A more techno-economic analysis is necessary for comparison between co-feed and redox mode for a given processing scheme.

Conclusion

Experimental results show the effects of pressure and steam on methane conversion and C_2+ selectivity with the OCM co-feed system and OCM chemical looping process. While pressure does increase methane conversion, C_2 selectivity decreases. Process simulations demonstrate that the OCM co-feed process for ethylene production is very energy intensive. The OCM reactor under redox mode is potentially advantageous with the recent development in chemical looping technologies. OCM chemical looping could lead to significant energy and cost reductions for the OCM process as an alternative technology for methane utilization.

References

1. Gesser, H.R., Hunter, N.R., Prakash, C.B. "The Direct Conversion of Methane to Methanol by Controlled Oxidation," *Chemical Reviews,* Volume 85, 1985

2. Keller, G.E., Bhasin, M.M. "Synthesis of Ethylene via Oxidative Coupling of Methane," Journal of Catalysis, Volume 73, 1982

3. Kondratenko, E.V., Baerns, M. "Oxidative Coupling of Methane," In: *Handbook of Heterogeneous Catalysis*; Ertl, G., Knozinger, H., Weitkamp, J., eds. Weinheim, Germany: Wiley-VCH; 2008.

4. Sofranko, J. A., Leonard, J. J, Jones, C. A. "The Oxidative Conversion of Methane to Higher Hydrocarbons," Journal of Catalysis, Volume 103, 1987

5. Jones, C. A., Leonard, J. J., Sofranko, J. A. "The Oxidative Conversion of Methane to Higher Hydrocarbons over Alkali-Promoted Mn/SiO_2," Journal of Catalysis. Volume 103, 1987

6. Li, F., Fan, L.-S. "Clean Coal Conversion Processes-Progress and Challenges," Energy and Environmental Science. Volume 1, 2008

7. Fan, L.-S. *Chemical Looping Systems for Fossil Energy Conversions*. New York, Wiley; 2010

8. Hinsen, W., Baerns, M. "Oxidative Coupling of Methane to C_2 Hydrocarbons in the Presence of Different Catalysts," *Chemiker-Zeitung*. Volume 107, 1983

9. Fang, X., Li, S., Gu, J., Yang, D. "Oxidative Coupling of Methane on W-Mn Catalysts," *Journal of Molecular Catalysis (China)*, Volume 6, NO. 6, 1992

10. Lee, Jong Y., Jeon, Wonjin, Choi, Jae-Wook et al. "Scaled-Up Production of C_2 Hydrocarbons by the Oxidative Coupling of Methane over Pelletized $Na_2WO_4/Mn/SiO_2$ Catalysts: Observing Hot Spots for the Selective Process," *Fuel*, Volume 106, 2013

11. Tiemersma, T.P., Tuinier, M.J., Gallucci, F., Kuipers, J.A.M., van Sint Annaland, M., "A Kinetics Study for the Oxidative Coupling of Methane on a $Mn/Na_2WO_4/SiO_2$ Catalyst," *Applied Catalysis A: General*, Volume 433-434, 2012

12. Dedov, A.G., Nipan, G.D., Loktev, A.S. et al. "Oxidative Coupling of Methane: Influence of the Phase Composition of Silica-Based Catalysts," *Applied Catalysis A: General*, Volume 407, 2011

13. Wang, J. Chou, L., Bing, Z. et al. "Comparative Study on Oxidation of Methane to Ethane and Ethylene over Na_2WO_4-Mn/SiO_2 Catalysts Prepared by Different Methods," *Journal of Molecular Catalysis A: Chemical*, Volume 245, 2006

14. Amenomiya, Y., Birss, V.I., Golezinowksi, M., Galuszka, J., Sanger, A.R. "Conversion of Methane by Oxidative Coupling," *Catalysis Review: Science and Engineering*, Volume 32, NO. 3, 1990

15. Hudgens, R.R., Silveston, P.L., Li, C.-Y., Adesina, A.A. Partial Oxidation and Dehydrogenation of Hydrocarbons. In *Periodic Operation of Reactors*; Silveston, P.L., Hudgens, R.R., Eds., Butterworth-Heinemann: Burlington, MA, 2013

16. Hugill, J.A., Tillemans, F.W.A., Dijkstra, J.W., Spoelstra, S. "Feasibility Study on the Co-generation of Ethylene and Electricity through Oxidative Coupling of Methane," *Applied Thermal Engineering*, Volume 25, NO. 8-9, 2005

17. Reay, D., Ramshaw, C., Harvey, A. Application Areas-Petrochemicals and Fine Chemicals. In *Process Intensification: Engineering for Efficiency, Sustainability and Flexibility*. Butterworth-Heinemann. Burlington, MA, 2008

18. Ghareghashi, A., Ghader, S., Hashemipour, H., Moghadam, H.R. "A Comparison of Co-current and Counter-current Modes for Fischer-Tropsch Synthesis in Two Consecutive Reactors of Oxidative Coupling of Methane and Fischer-Tropsch," *Journal of Natural Gas Science and Engineering*, Volume 14, 2013

19. Graf, P. Combining Oxidative Coupling and Reforming of Methane. Vision or Utopia? Ph.D. Dissertation, University of Twente, Enschede, The Netherlands, 2008

20. Tiemersma, T.P., Kolkman, T., Kuipers, J.A.M., van Sint Annaland, M. "A Novel Autothermal Reactor Concept for Thermal Coupling of the Exothermic Oxidative Coupling and Endothermic Steam Reforming of Methane," *Chemical Engineering Journal*, Volume 203, 2012

21. Salerno, D., Arellano-Garcia, H., Wozny, G.. "Techno-Economic Analysis For Ethylene and Oxygenates Products from the Oxidative Coupling of Methane," *Chemical Engineering Transactions*, Volume 24, 2011

22. Salerno, D. Optimal Synthesis of Downstream Processes using the Oxidative Coupling of Methane Reaction. Ph.D. Dissertation, Technical University of Berlin, Berlin, Germany, 2013

23. "Ocmol Website Home," http://www.ocmol.eu/index.php (accessed April 10, 2015)

24. Ahari, J.S., Sadeghi, M.T., Pashne, S.Z. "Optimization of OCM Reaction Conditions over Na-W-Mn/SiO_2 Catalyst at Elevated Pressure," *Journal of the Taiwan Institute of Chemical Engineers*, Volume 42, 2011

25. Karimi, A., Ahmadi, R., Bozorg Zadeh, H.R., Jolodar, J.A., Barkhordarion, A. "Catalytic Oxidative Coupling of Methane-Experimental Investigation and Optimization of Operational Conditions," *Petroleum & Coal*, Volume 49, NO. 3, 2007

26. Gaffney, A.M. Methane Conversion Process. U.S. Patent 4,788,372, November 29, 1988

27. Ahmad, N. "TASNEE Ethylene Cracker Plant: CGC Inter Stage Cooler Fouling and its Remedial Approach to Avoid Unplanned Shutdown," Talk presented at: Ethylene Middle East Technology Conference; 19-21 October, 2014, http://ethylene-me.com/forum/presentations/CGC%20Inter%20Stage%20Cooler%20Fouling_Nafees%20Ahmad_TASNEE.ppsx (accessed April 12, 2015)

400048 Synthesis of Tungsten Carbide Nanoparticles in Biochar Matrix As a Catalyst for Dry Reforming of Methane to Syngas

Abstract:

Tungsten carbides nanoparticles were synthesized by carbothermal reduction (CR) of tungsten promoted biochar. The nanoparticles were characterized for physicochemical properties by multiple morphological and structural methods (e.g. SEM, TEM, and XRD). Characterization results revealed that the tungsten carbide nanoparticles formation involves the sequence step: $WO_3 \rightarrow WO_2 \rightarrow W \rightarrow W_2C \rightarrow WC$. The catalytic performance for dry reforming of methane over these tungsten carbide nanoparticles was investigated. The stability testing result of the tungsten carbide nanoparticles shows no catalyst deactivation for the duration of the 500 hours.

15 AIChE Spring Meeting, Austin, TX

Solar Thermochemical Reforming of Natural Gas via Metal Oxide Based Redox Reactions

(Principal Author)
Rahul Bhosale
Department of Chemical Engineering
College of Engineering
Qatar University, Doha, Qatar
rahul.bhosale@qu.edu.qa

(Other Authors)
Anand Kumar
Department of Chemical Engineering
College of Engineering
Qatar University, Doha, Qatar
akumar@qu.edu.qa

Fares A. AlMomani
Department of Chemical Engineering
College of Engineering
Qatar University, Doha, Qatar
falmomani@qu.edu.qa

Shahd Gharbia
Department of Chemical Engineering
College of Engineering
Qatar University, Doha, Qatar
dd1202223@student.qu.edu.qa

Dareen Dardor
Department of Chemical Engineering
College of Engineering
Qatar University, Doha, Qatar
sg1000354@student.qu.edu.qa

Moustafa Hussein Ali
Department of Chemical Engineering
College of Engineering
Qatar University, Doha, Qatar
ma1203561@student.qu.edu.qa

Jamila Folady
Department of Chemical Engineering
College of Engineering
Qatar University, Doha, Qatar
jf1003686@student.qu.edu.qa

Shiva Yousefi
Department of Chemical Engineering
College of Engineering
Qatar University, Doha, Qatar
sy1106491@student.qu.edu.qa

Mehak Jilani
Department of Chemical Engineering
College of Engineering
Qatar University, Doha, Qatar
mj1104333@student.qu.edu.qa

Prepared for Presentation at
American Institute of Chemical Engineers
2015 Spring Meeting
11th Global Congress on Process Safety
Austin, Texas
April 27-30, 2015

UNPUBLISHED

Solar Thermochemical Reforming of Natural Gas via Metal Oxide Based Redox Reactions

Rahul Bhosale
Department of Chemical Engineering
College of Engineering
Qatar University, Doha, Qatar
rahul.bhosale@qu.edu.qa

Anand Kumar, Fares A. AlMomani, Shahd Gharbia, Dareen Dardor, Moustafa Hussein Ali, Jamila Folady, Shiva Yousefi, Mehak Jilani

Department of Chemical Engineering
College of Engineering
Qatar University, Doha, Qatar

Keywords: Samarium Sesquioxide, Methane Reforming, Solar Syngas, Computational Thermodynamic Modeling

Abstract

This paper reports the thermodynamic analysis of Sm_2O_3 based solar thermochemical CH_4 reforming process. Thermodynamic equilibrium compositions of the syngas production via CH_4 reforming via Sm_2O_3 reduction as a function of reaction temperature were determined. Second law thermodynamic analysis was performed to determine the cycle efficiency and solar to fuel energy conversion efficiency attainable without heat recuperation.

1. Introduction

Several thermochemical cycles such as iron oxide cycle [1-6], zinc/zinc oxide cycle [7-10], tin/tin oxide cycle [8, 11-13], mixed ferrite cycle [14-21], ceria cycle [22-24], sulfur-iodine cycle [25-27], and hybrid sulfur cycle [28-29] were investigated towards water splitting or CO_2 splitting reactions to produce solar syngas. However, lower solar to fuel conversion efficiency is one of the drawbacks of these cycles. The reason for the lower efficiency is the unnecessary heat loss during the heating of inert Ar.

Therefore, in this paper, we propose solar thermochemical reforming of CH_4 via metal oxide reduction (samarium sesquioxide) for the production of syngas without using the inert gas. In this study, the thermodynamic feasibility of this cycle was investigated and obtained results are presented in detail. Thermodynamic equibrium composition and the required reforming temperature was identified for m_2O_3 based solar thermochemical CH_4 reforming process. Also,

Note: Do not add page numbers. Do not refer to page numbers when referencing different portions of the paper

the maximum theoretical solar energy conversion efficiency of the process is determined by performing the second law thermodynamic analysis.

2. Chemical Thermodynamic Equilibrium

2.1 Solar thermochemical reforming of CH_4 using Samarium Sesquioxide

Figure 1 represents the thermodynamic equibrium compositions for the solar thermochemical reforming of CH_4 using Sm_2O_3. Species with concentrations below 2×10^{-2} moles were neglected. The temperature of the reforming was determined once 2 moles of H_2 and 1 mole of CO is produced via CH_4 reforming.

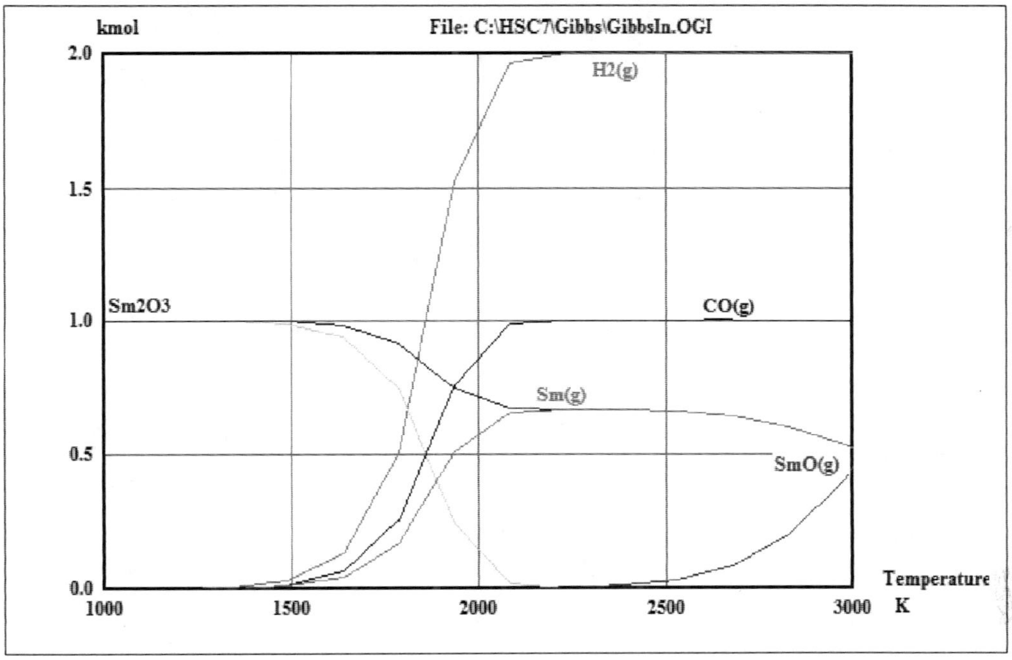

Figure 1. Solar thermochemical reforming of CH_4 using Sm_2O_3 (Ar = 0 moles/sec).

According to the obtained results, it was realized that the production of high quality syngas (H_2/CO = 2:1) via Sm_2O_3 based solar thermochemical CH_4 reforming is possible at 2200 K. Also, following equation represents the reaction mechanism for the Sm_2O_3 based solar thermochemical CH_4 reforming process.

$$Sm_2O_3 + CH_4(g) \rightarrow 0.656Sm_2O_3 + 0.657Sm(g) + 2H_2(g) + CO(g) + 0.031SmO(g) \quad (1)$$

3. Exergy Analysis

The solar to fuel conversion efficiency and exergy analysis of the Sm_2O_3 based solar thermochemical CH_4 reforming process was carried out by following the methodology and governing equations derived and used previously for other MO based solar thermochemical cycles [7]. Thermodynamic properties were extracted from HSC software and databases and the analysis is normalized to the Sm_2O_3 molar flow rate (1 mol/sec) entering the solar reactor.

The process flow diagram for the Sm_2O_3 based solar thermochemical CH_4 reforming process is presented in Figure 2. This process includes a solar reactor, theoretical H_2/O_2 ideal fuel cell, and a quench. Further assumptions such as a) The solar reactor is assumed to be a perfectly insulated blackbody absorber, b) No convective/conductive heat losses, c) Effective emissivity and absorptivity equal to unity, d) All the products separate naturally without expending any work are also made for the simplification of the process.

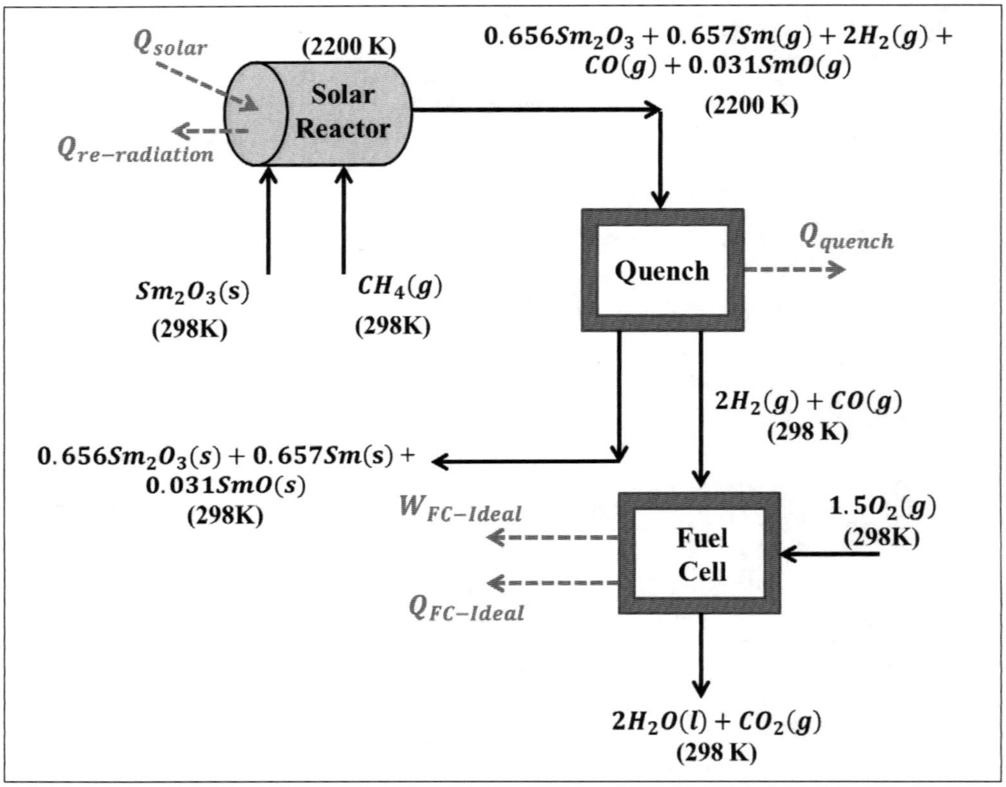

Figure 2. Process configuration for Sm_2O_3 based solar thermochemical CH_4 reforming process.

3.1 Solar energy absorption efficiency ($\eta_{absorption}$)

The solar energy absorption efficiency ($\eta_{absorption}$) is defined as the net rate at which energy is being absorbed by the solar reactor divided by the solar power input through the aperture of the solar reactor. The $\eta_{absorption}$ can be calculated as:

$$\eta_{absorption} = 1 - \left(\frac{\sigma T_R^4}{IC}\right) \qquad (2)$$

Where, σ = Stefan-Boltzmann constant, I = normal beam insolation $1(kW/m^2)$, C= solar flux concentration ratio of the solar concentrating system (5000 suns), $Q_{reactor-net}$= net energy absorbed in the solar reactor, Q_{solar} = solar energy input. For Sm_2O_3 based solar thermochemical CH_4 reforming process, at T_R= 2200 K, the $\eta_{absorption}$ is observed to be equal to 73.43%.

3.2 *Net solar energy absorbed*

For the operation of Sm_2O_3 based solar thermochemical CH_4 reforming process, the energy required to perform the reforming of CH_4 using Sm_2O_3 ($Q_{reactor-net}$) is needs to be fulfilled by solar energy input. As per the thermodynamic calculations, the $Q_{reactor-net}$ for this process is equal to 1250 kW.

3.3 *Solar energy input to the solar reactor* (Q_{solar})

Based on the calculated $\eta_{absorption}$ and $Q_{reactor-net}$, the amount of solar energy input required to operate the solar reactor (Q_{solar}) dealing with the Sm_2O_3 based solar thermochemical CH_4 reforming process can be determined by equation (3) and observed to be equal to 1702 kW.

$$Q_{solar} = \frac{Q_{reactor-net}}{\eta_{absorption}} \qquad (3)$$

3.4 *Radiation heat losses from the solar reactor*

Radiation heat losses and % re-radiation from the solar reactor conducting the Sm_2O_3 based solar thermochemical CH_4 reforming process can be calculated by equations (4) and (5):

$$Q_{re-radiation} = Q_{solar} - Q_{reactor-net} \qquad (4)$$

$$\%\, re - radiation\ losses = \frac{Q_{re-radiation}}{Q_{solar}} \times 100 \qquad (5)$$

Computational thermodynamic modeling indicate that radiation heat losses = 452 kW and % re-radiation = 26.56%.

3.5 *Rate of heat rejected to the surrounding due to quenching*

To avoid the recombination, the products coming out of the solar reactor were quenched immediately to the ambient temperature (298K) by using a quench and the heat rejected by the quench is calculated by Eq. (6) which is equal to 660 kW.

$$Q_{quench} = -\dot{n}\Delta H|_{\substack{0.656Sm_2O_3(s)+0.657Sm(g)+2H_2(g)+CO(g)+0.031SmO(g)@2200K\rightarrow \\ 0.656Sm_2O_3(s)+0.657Sm(s)+2H_2(g)+CO(g)+0.031SmO(s)@298K}} \qquad (6)$$

3.6 *Rate of theoretical work performed and heat rejected by the ideal fuel cell*

Sm_2O_3 based solar thermochemical CH_4 reforming process is completed with the addition of an ideal $H_2/CO/O_2$ fuel cell with 100% work efficiency (assumption). This ideal fuel cell is added to the cycle to extract the maximum work from the net syngas produced by this process. The rate of theoretical work performed by the ideal fuel cell is calculate by equation (7) and observed to be equal to 730.25 kW. Likewise, the rate of heat energy released by the ideal fuel cell is determined as equal to 114.864 kW [according to equation (8)].

$$W_{FC-Ideal} = -\dot{n}\,\Delta G|_{2H_2(g)+CO(g)+1.5O_2(g)@298K \rightarrow 2H_2O(l)+CO_2(g)@298K} \quad (7)$$

$$Q_{FC-Ideal} = -(298) \times \dot{n}\,\Delta S|_{2H_2(g)+CO(g)+1.5O_2(g)@298K \rightarrow 2H_2O(l)+CO_2(g)@298K} \quad (8)$$

3.7 Solar to fuel conversion efficiency

Cycle efficiency of the Sm_2O_3 based solar thermochemical CH_4 reforming process is defined as the ratio of theoretical work performed by the ideal fuel cell to the solar energy input:

$$\eta_{cycle} = \frac{W_{FC-Ideal}}{Q_{solar}} \quad (9)$$

Furthermore, the solar-to-fuel energy conversion efficiency of the Sm_2O_3 based solar thermochemical CH_4 reforming process is defined as the ratio of higher heating value (HHV) of the syngas produced to the solar energy input:

$$\eta_{solar-to-fuel} = \frac{HHV_{syngas}}{Q_{solar}} \quad (10)$$

Where,

$$HHV_{syngas} = -\dot{n}\,\Delta H|_{2H_2(g)+CO(g)+1.5O_2(g)@298K \rightarrow 2H_2O(l)+CO_2(g)@298K} \quad (11)$$

As per the HSC computation thermodynamic modeling, η_{cycle} and $\eta_{solar-to-fuel}$ equal to 42.90% and 50.21%, respectively.

4. Conclusion

Sm_2O_3 based solar thermochemical CH_4 reforming process have been thermodynamically investigated. Thermodynamic equilibrium calculations performed at 1 atm pressure and in absence of inert Ar indicate that the CH_4 reforming via Sm_2O_3 reduction is feasible at 2200K. The second law thermodynamic analysis indicate that indicate $_{absorption}$ = 74.43%, Q_{solar}= 1702 kW, $Q_{reactor-net}$ = 1250 kW, $Q_{re-radiation}$ = 452 kW, Q_{quench}= 660 kW, $W_{FC-Ideal}$=730.25 kW, $Q_{FC-Ideal}$= 114.864 kW, η_{cycle}= 42.90%, and $\eta_{solar-to-fuel}$= 50.21%, respectively.

5. Acknowledgments

The authors gratefully acknowledge the financial support provided by the Qatar University Internal Grants QUUG-CENG-CHE-13/14-4 and QUUG-CENG-CHE-14\15-10.

References

[1] Agrafiotis C, Pagkoura C, Lorentzou S, Kostoglou M, Konstandopoulos AG. Hydrogen production in solar reactors. Catalysis Today 2007;127:265–77.

[2] Steinfeld A, Sanders S, Palumbo R. Design aspects of solar thermochemical engineering – A case study: two-step water-splitting cycle using the Fe3O4/FeO redox system. Solar Energy 1999;43:43–53.

[3] Kodama T, Nakamuro Y, Mizuno T. A two-step thermochemical water spitting by iron-oxide on stabilized zirconia. Journal of Solar Energy Engineering 2006;128:3–7.

[4] Gokon N, Murayama H, Umeda J, Hatamachi T, Kodama T. Monoclinic zirconia-supported Fe3O4 for the two-step water-splitting thermochemical cycle at high thermal reduction temperatures of 1400–1600°C. International Journal of Hydrogen Energy 2009;34:1208–17.

[5] Ishihara H, Kaneko H, Hasegawa N, Tamaura Y. Two-step water-splitting at 1273–1623K using yttria-stabilized zirconia-iron oxide solid solution via co-precipitation and solid-state reaction. Energy 2008;33:1788–93.

[6] Miller J, Allendorf M, Diver R, Evans L, Seigel N, Stuecker J. Metal oxide composites and structures for ultra-high temperature solar thermochemical cycles. Journal of Material Science 2008;43:4714–28.

[7] Steinfeld A. Solar hydrogen production via a two-step water-splitting thermochemical cycle based on Zn/ZnO redox reactions. International Journal of Hydrogen Energy 2002;37:611–19.

[8] Abanades S. Thermogravimetric analysis of CO2 and H2O reduction from solar nanosized Zn powder for thermochemical fuel production. Industrial & Engineering Chemistry Research 2012;51:741–50.

[9] Ernst F, Steinfeld A, Pratsinis S. Hydrolysis rate of submicron Zn particles for solar H2 synthesis. International Journal of Hydrogen Energy 2009;34:1166–75.

[10] Vishnevetsky I, Epstein M. Production of hydrogen from solar zinc in steam atmosphere. International Journal of Hydrogen Energy 2007;32:2791–2802.

[11] Abanades S, Charvin P, Lemont F, Flamant G. Novel two-step SnO2/SnO water-splitting cycle for solar thermochemical production of hydrogen. International Journal of Hydrogen Energy 2008;33:6021–30.

[12] Charvin P, Abanades S, Lemont F, Flamant G. Experimental study of SnO2/SnO/Sn thermochemical systems for solar production of hydrogen. AIChE Journal 2008;54:2759–67.

[13] Vishnevetsky I, Epstein M. Tin as a possible candidate for solar thermochemical redox process for hydrogen production. Journal of Solar Energy Engineering 2009;131:021007-1–021007-8.

[14] Scheffe JR, Li J, Weimer AW, A spinel ferrite/hercynite water-splitting redox cycle. International Journal of Hydrogen Energy 2010;35:3333–3340.

[15] Bhosale RR, Shende RV, Puszynski JA, H2 generation from thermochemical water-splitting using sol-gel derived Ni-ferrite. Journal of Energy and Power Engineering 2010;4:27–38.

[16] Neises M, Roeb M, Schmucker M, Sattler C, Pitz-Paal R. Kinetic investigations of the hydrogen production step of a thermochemical cycle using mixed iron oxides coated on ceramic substrates. International Journal of Energy Research 2010;34:651–61.

[17] Bhosale RR, Shende RV, Puszynski JA, H2 generation from thermochemical water-splitting using sol-gel synthesized Zn/Sn/Mn-doped Ni-ferrite. International Review of Chemical Engineering 2010;2:852–62.

[18] Lorentzou S, Zygogianni A, Tousimi K, Agrafiotis C, Konstandopoulos AG. Advanced synthesis of nanostructured material for environmental applications. Journal of Alloys and Compounds 2009;483:302–05.

[19] Bhosale RR, Khadka RP, Shende RV, Puszynski JA. H2 generation from two-step thermochemical water-splitting reaction using sol-gel derived SnxFeyOz. Journal of Renewable and Sustainable Energy 2011;3:063104-1–063104-16.

[20] Varsano F, Padella F, Alvani C, Bellusci M, Barbera A. Chemical aspects of the water-splitting thermochemical cycle based on sodium manganese ferrite. International Journal of Hydrogen Energy 2012;37:11595–601.

[21] Bhosale RR, Shende RV, Puszynski JA. Thermochemical water-splitting for H2 generation using sol-gel derived Mn-ferite in a packed bed reactor. International Journal of Hydrogen Energy 2012;37:2924–34.

[22] Scheffe JR, Steinfeld A. Thermodynamic analysis of cerium-based oxides for solar thermochemical fuel production. Energy & Fuels 2012;26:1928–36.

[23] Abanades S, Flamant G. Thermochemical hydrogen production from a two-step solar-driven water-splitting cycle based on cerium oxides. Solar Energy 2006;80:1611–23.

[24] Gal A, Abanades S. Dopant incorporation in ceria for enhanced water-splitting activity during solar thermochemical hydrogen generation. Journal of Physical Chemistry C 2012;116:13516–523.

[25] Lin X, Zhang Y, Wang Z, Wang R, Zhou J, Cen K. Hydrogen production by HI decomposition over nickel–ceria–zirconia catalysts via the sulfur–iodine thermochemical water-splitting cycle. Energy Conversion and Management 2014;84:664–70.

[26] Murphy IV J, O'Connell J. Process simulations of HI decomposition via reactive distillation in the sulphur-iodine cycle for hydrogen manufacture. International Journal of Hydrogen Energy 2012;37:4002–11.

[27] Dehghani S, Sayyaadi H. Energy and exergetic evaluations of Bunsen section of the sulfur–iodine thermochemical hydrogen production plant. International Journal of Hydrogen Energy 2013;38:9074–84.

[28] Lin S, Flaherty R. Design studies of the sulphur trioxide decomposition reactor for the sulphur cycle hydrogen production process. International Journal of Hydrogen Energy 1983;8:589–96.

[29] Hinkley J, O'Brein J, Fell C, Lindquist S. Prospects for solar only operation of the hybrid sulphur cycle for hydrogen production. International Journal of Hydrogen Energy 2011;8:589–96.

AUTHOR INDEX

AUTHOR INDEX

American Institute of Chemical Engineers
3 Park Avenue
New York, New York 10016-5991

CURRAN ASSOCIATES INC
proceedings.com